MW01349844

NAPOLEONIC WARGAMING
FOR FUN

NAPOLEONIC WARGAMING FOR FUN

Paddy Griffith

WARD LOCK LIMITED · LONDON

To my parents

First published in Great Britain in 1980
by Ward Lock Limited, 116 Baker Street,
London W1M 2BB, a Pentos Company.

Designed by Charlotte Westbrook
House editor Gill Freeman
Text filmset in 11 point Apollo
by Servis Filmsetting Limited,
Manchester

Printed in Great Britain by
J.W. Arrowsmith Ltd, Bristol
Bound by Waterlow & Sons, Dunstable

**British Library Cataloguing in
Publication Data**

Griffith, Paddy
Napoleonic wargaming.
1. War games 2. Europe – History – 1789–1815
I. Title
793'.9 U310

ISBN 0-7063-5813-9
0-7063-6042-7 Pbk

Frontispiece *A rifleman of the 95th. This regiment took a leading part in the British skirmish line during the Peninsular Wars.*

Contents

Acknowledgements

This book could not have been written without the encouragement and forbearance, over the years, of my wargaming friends. They have given me not only many good games, but also many good ideas and pieces of advice. I would especially like to thank Peter McManus for his help with map kriegspiels and TEWTS; John Davis for the decisive impetus he gave to the army level game; Henrik Kiertzner for his enthusiastic skirmishing; and Andy Callan for his sustained inspiration in the brigade and generalship games. I am also grateful to Nigel de Lee, as well as to past and present members of the Lancaster University Jomini Group, who have helped me on many occasions. Finally, I would like to thank Christopher Duffy for demonstrating – on the 'night of the Puffer' – the correct approach to publishing wargames.

The author and publishers would like to thank the following for kindly supplying photographs for this book:
Model and Allied Publications, pages 10, 19
The Mansell Collection, pages 33, 57, 62, 86
The frontispiece and the photograph on page 28 were taken by Rob Matheson, and the painted figures were lent by Seagull Models (GB) Limited, 32 Thurloe Place, London SW7.
The photographs on pages 38, 52, 53, 54 and 65 were taken by Christopher Duffy.

Line illustrations drawn by Design Practitioners Limited

1
Basic ideas

This is a book about Napoleonic wargames; games which people can play between themselves for amusement first and foremost, but also to deepen their interest in the Napoleonic Wars. It will not require any previous knowledge or experience from the reader, and will try to provide full explanations as it goes along – a little about Napoleonic battles and much about the wargames. It does, however, aim to be more than simply a manual for beginners, since well over half of its pages are concerned with the 'higher art' of Napoleonic wargaming. If you are a beginner, therefore, you should tarry long over the early sections, and then graduate slowly and cautiously into the later parts. If you are an experienced Napoleonic wargamer, on the other hand, you could skip quickly through the first few chapters, and come rapidly to grips with the more advanced matter later on. You may agree or disagree with the details of my ideas; but I sincerely hope that at least their general structure will give you something to work on and develop, according to your individual taste.

No one can be very dogmatic about wargame rules, because they are always a highly personal thing. Every club or small group of wargamers will understand any specific set of rules in a rather different way from everyone else, and every player will add his own pet interpretations and short cuts. This individual tinkering with the rules is, to my mind, one of the great strengths of the hobby. Every wargamer can twist the other fellow's ideas to suit his own particular approach, so that the result is a personal statement of what Napoleonic warfare represents to him. There is no higher authority to tell him that he is wrong apart from the Delphic, ambiguous, and often unreliable statements of the history books. This present collection of wargame rules is therefore intended to be a record of my own personal thinking. It cannot be 'definitive'; and readers are cordially invited to change round, or ignore, any of my suggestions which don't quite suit their own views.

You will find here a selection of the Napoleonic wargames which I have developed over the years, together with some of the general wargaming ideas which lie behind them. I start with two fairly conventional games for beginners, and then branch out to explore a variety of more advanced games for the experienced player. I do not confine myself to presenting a single system of play: instead, I aim to offer as many different approaches as possible, so that the reader can decide for himself which ones he likes best.

This wide diversity may sound surprising to many wargamers, because there are usually only three types of Napoleonic game in general use. These are:

Skirmish games, which portray very low level action indeed, where one model soldier represents one real man.

Divisional games, where several hundred model soldiers represent a force of around Divisional strength. This usually works out as one model soldier to thirty-three real men.

Board games, without model soldiers at all, but with units represented by cardboard markers moving on a gridded map. The level of action may vary from a battalion to a group of armies.

Of these three categories I do not intend to discuss the third – board wargaming – at all, since it already has its own extensive literature (see especially N. Palmer's *The Comprehensive Guide to Board Wargaming,* London, 1977), and it is really a quite distinct activity from other types of wargames. Normally, therefore, we would be left with a rather narrow choice between skirmish and Divisional games. In the present book, however, I shall extend this choice to include several additional types of Napoleonic game.

The brigade game, half way between the skirmish and Divisional games. In this case each model soldier represents about ten men, instead of either one or thirty-three. This allows entire battalions to be represented, but in sufficient detail to provide some very low level tactical problems.

The army level game, in which only very high level units are represented. This allows the full grandiose sweep of Napoleonic battle tactics to be played, albeit without much attention to lower level tactics.

It may be that the sheer scale of the army level game makes the whole thing seem rather too abstract and impersonal. In this case we can redress the balance by playing the *generalship game.* This is a distant relative of board games, and does not use model soldiers. It is set at the level of grand strategy, but seen from a very personal point of view. The player has to imagine himself as an all-round leader on campaign; not simply a specialized tactician or staff officer.

All of the above levels of play can be approached through a rather different style of game: the *map kriegspiel.* This is not, perhaps, a very new idea; but it has recently been sadly neglected by wargamers.

A further alternative (which may also be applied to most levels of play) is the *Tactical exercise without troops.* This is a way of taking the wargame out of doors.

Because of the wide variety of different games available, the player can select exactly the right one to suit his particular interest at a given moment. If he has just read a book about British riflemen in the outpost line during the Peninsular War, for example, he will be able to re-create their exploits with the skirmish game. If he has been reading a biography of Wellington on campaign, on the other hand, then he will want to try the generalship game. If battalion tactics (for example, volley firing, forming square, making bayonet charges, etc.) interest him, he has the brigade game. If army battle tactics (for example, conserving a *masse de rupture*, co-ordinating flanking corps, choosing an order of march, etc.) attract him, the army level game is tailor-made. By using a variety of different approaches the player can highlight a variety of different aspects of the Napoleonic Wars. He will not be stuck in the same old (Divisional) rut every time.

Realism and pretence

An important point to remember with all types of wargame is that they are fundamentally different from formal games like chess or bridge. With the latter, the fun of the game is derived from the purely abstract competition between the two players. It is the ability to think logically, and almost mathematically, which is most important. In a wargame, on the other hand, the competitive element certainly has a part to play; but it is not really predominant. It is, rather, the sense of re-creating the past which provides the excitement. The wargamer is playing at 'let's pretend' with military history. Some form of realism, or imaginative leap into the Napoleonic era will be essential and anything which helps it along will be good for the wargame.

There are three main areas in which it is practical to hope for realism in wargames. These are:

Aesthetic We can use finely detailed model soldiers and scenery to make a sort of animated miniature *tableau* of the scene. This in itself can sometimes bring the event vividly to life, and four of the games in this book are based upon it.

Tactical We can find out the vital statistics of regimental fighting during the Napoleonic period, and make sure that the rules of our game are tactically realistic. In other words we will make our model regiments march, shoot and fight according to the same limitations and probabilities as were really imposed upon Napoleonic regiments. To do this we will need to know some accurate details of Napoleonic tactics, and then to use various wargame techniques to fit them into a workable set of rules.

Command and control It is all very well knowing how Napoleonic regiments operated; but this is really only half the story. In the more advanced games in this book we will also try to develop rules for the other half: the ways Napoleonic commanders operated. Regimental action was certainly an important instrument by which commanders

A far cry from chess. In this wargame a very high level of aesthetic realism has been achieved, in both the terrain features and the finely-detailed model soldiers.

could win victories, but it was by no means the only one. Commanders also had to master a whole set of techniques for collecting intelligence, issuing orders, and controlling the battle. If we can make a game which simulates these things, and forces us to make the same types of decisions as Napoleonic commanders would really have had to make, then we will have gone a long way towards imagining ourselves in their shoes. We will no longer simply be chess players sitting over our tables and moving pieces when and where we wish. Instead, we will be commanders in battle who have to cope with misleading information, reluctant subordinates, couriers who get lost, and a host of other difficulties before we can get our pieces into motion at all.

The realism of our rules – the sense that we are really taking the part of a Napoleonic commander – will be easier to achieve if we read novels, biographies, and history books about the period. The more we know about our subject, the more we will be able to look at it in the right perspective, and imagine ourselves in the middle of it all. It is for this reason that I have included a short book list for each game.

It is also extremely important to prevent our rules from becoming over-complex, thus sacrificing playability to realism. Napoleonic commanders did not have to be mathematical wizards to win their wars; nor did barrack-room lawyers often rise to high command. We must therefore steer well clear of pedantry, petty detail, and rule-mongering in our games and concentrate firmly upon general command decision-making.

Many wargamers fail to achieve the right balance in all this, and either become obsessed with minor details and mental arithmetic in the name of realism, or go to the opposite extreme, and reject any claim to realism at all, in the name of playability. Neither of these approaches is correct, since it is perfectly possible to devise games which are both realistic and playable. To do this, however, you must be clear about just what level of realism you are after; through exactly whose eyes you are trying to look at the battle.

The games in the present book have varying standards of realism and playability, but they all try to be very clear about just what is being simulated, and precisely whom the player is supposed to be. In this way they attempt to combine the two elements of fun and historical satisfaction; although at the end of the day the reader will of course only get pleasure from them in proportion to the time and energy he himself puts in. I have at least tried to keep the games relatively short, so that their play-mechanics may be kept to a minimum: each game is designed to be completed within two to four hours, by two to six people.

Table of units and formations normally used in the Napoleonic Wars: (note that each nation had a different specific organization of its own; the details given here are only a very rough guide)

Unit/formation	*Rank of commander*	*Approx. strength*	*Average number and type of sub-units*
Army	Sovereign or marshal	40–300,000	2–12 × Army corps
Army corps	Marshal or general	10–50,000; about 50 guns	2–5 × Division
Division	General 'of division'	2–12,000; 6–24 guns	2–5 × Brigade *or* × regiment
Brigade	General 'of brigade'	1–3,000	2–5 Regiment
Regiment	Colonel	1–3,000	1–5 × Battalion
Battalion	Lt. col/major/ commandant	200–1,200	4–10 × Company
Company	Captain	50–150	2 × Section

2
The skirmish game

The first game we meet is classic in its simplicity, and therefore a very good place for learners to start. It is the skirmish game, where each model soldier represents one real man, so there are no complexities in the scales. Also, owing to the small size of the action, we do not portray the rather technical evolutions of battalions and regiments. Instead, we must content ourselves with the relatively straightforward tactics of small groups. The level of action is somewhere between the individual soldier and the section of forty men or so.

The types of troops in Napoleonic skirmishes

The troops engaging in Napoleonic skirmishes could vary enormously in type, and might be drawn from almost any branch of the service. Each army had its own specialized formations, and it is impossible here to give a comprehensive listing of them all. Nevertheless, as a help to the beginner I shall mention some of them.

Line infantry

This was the normal regular infantry, often composed of conscripts who had not been given a long military training, but who were under the eye of a *cadre* of veteran N.C.O.s and officers. These troops were of patchy quality but, on the whole, quite reasonable. They were armed with a muzzle-loading flintlock musket and bayonet, up to sixty rounds of ammunition, and perhaps a short sabre or machete. Battalions were composed of a number of companies (the precise number varied in each army), each of two sections. The sections were in turn composed of squads of eight to ten men.

Companies might contain a total of eighty men or more, although after a little time on campaign this figure would usually be very seriously reduced. As part of the company total there would also be a number of officers and N.C.O.s: for example, the British regulations called for three officers, three sergeants, four corporals, and one drummer. The French had rather more, and were very proud of the greater supervision and leadership which this supposedly bestowed upon their armies.

Most companies were designated centre or fusilier companies, and consisted of the ordinary rank and file. In each battalion, however, there were also, usually, two flank or élite companies, composed of the better, or longest serving soldiers. One of these flank companies would consist of light infantry; the other of grenadiers. In theory the light infantry was supposed to be sent forward to skirmish and screen the battalion's front, while the grenadiers were supposed to spearhead attacks or form a reserve rallying point. In practice, however, these rules seem to have been broken quite as much as they were observed. Precisely because they were made up of the best soldiers, both flank companies were often sent forward indiscriminately on any hazardous duty, regardless of whether it was technically a 'light' or a 'grenadier' task. In our skirmishes, therefore, we might well expect to see rather a lot of the light infantry companies, but by no means on every occasion.

Militia or Landwehr

Many countries used second line troops in the field to make up numbers. They tended to be badly trained and led, possibly under- or over-age, and despised by other troops. In action, nevertheless, they often performed surprisingly well. Their organization was similar to that of the line battalions.

Irregulars

Irregulars were often used when nothing better was available. These troops would have extremely sketchy organization, and probably very poor equipment. In formal battles their units would be hopelessly fragile; but they performed better in smaller skirmishes or guerrilla raids, where enthusiasm and individual cunning counted for more than formal drill. We may thus have an amusing time designing irregular bands for our games, and arming them with such diverse weapons as nail guns, blunderbusses, billhooks, and stilettoes.

Guards

Any type of unit – infantry, cavalry, artillery, and even military police or marines – could be designated as Guards, i.e., the personal troops of the sovereign. In battle they would usually be kept back as the last reserve, 'so as not to spoil them'. They enjoyed enormous privileges, and according to contemporary wisdom they were supposed to be unbeatable, except perhaps by the enemy's Guards. In the light of Second World War experience, however, this assumption may perhaps be questioned. Relatively fresh units, provided they have done well in their first battles, often tend to fight better than veteran troops who have been through the crucible too often.

Light infantry regiments

Most nations designated certain whole regiments as light infantry. In most cases, particularly in the French army, this tag was absolutely

Some Napoleonic troop types: from left British hussar, British rifleman, Cossack, French Guard infantry, French dragoon, Russian line infantry.

meaningless. The light infantry regiments were identical to the line, apart from the difference in nomenclature and uniforms. Both line and light regiments would be used indiscriminately for skirmishing. In some other armies, however, light infantry really did perform a specialist role, as highly trained skirmishers and sharpshooters. This was particularly true of units armed with the rifle, e.g., the British rifle regiments, or the Tyrolean Jaeger. These troops not only possessed superior weapons to most of their counterparts, but they also enjoyed the rare distinction of knowing how to use them.

Heavy cavalry

The battle cavalry of the Napoleonic Wars included armoured (cuirassiers, carabiniers, etc.) and unarmoured types (heavy dragoons, horse grenadiers, etc.). Medium cavalry (dragoons, lancers, etc.) was also normally used in this role, so for convenience, I shall call all these types heavy cavalry, with the single exception of the lancers. All the heavy cavalry tended to consist of 'big men on big horses', and was trained specifically for massed shock action. They did skirmish from time to time, especially the dragoons, but normally they preferred to leave that sort of thing to the light cavalry.

All cavalry was armed with a sword (usually curved for the light cavalry, straight for the heavy) and a small musket or carbine. Heavier types often had pistols as well, while dragoons would carry bayonets in case they were ordered to fight on foot. The basic organization was the regiment, which consisted of a number of squadrons, each of two platoons, of two sections each. The squadron might have a total of 120 men, although on campaign this figure would dwindle even faster than in the infantry. Good horse care was essential to keep up the fighting strength of the cavalry, and it took up much of their time.

Light cavalry

There were several different types of light cavalry (light dragoons, hussars, chasseurs, cheveau-legers, etc.) to which for the purposes of this book we shall add the (technically medium) lancers. The primary role of all these troops was scouting, liaison and outpost work; although in practice they tended to be used as battle cavalry more often than they should have been. They rode small horses, and felt especially comfortable when they were skirmishing.

Irregular cavalry

In a few armies there was an unsophisticated toleration of irregular cavalry, and in Egypt the Mameluke army actually contained nothing else. As far as Europe was concerned, the Cossacks formed the most famous and the most feared irregular cavalry force, although other examples could be cited from Spain or Prussia. This type of soldier is of especial interest to us here, since skirmishing was his only means of livelihood. As with irregular infantry, he might be armed with some very original weapons, ranging from the very long Cossack lances to the short bows of the Bashkirs.

Artillery

Artillery would rarely take part in skirmish actions, although raiding cavalry might carry the occasional light piece with it; or guerrillas might lay ambushes for artillery convoys on the march. To complete our list of troop types, however, it is worth saying a few words about what was, after all, probably the most powerful weapon of the Napoleonic Wars.

Strictly speaking, all guns on mobile carriages (as opposed to coastal, naval, fortress or siege artillery) should be termed field artillery. For the purposes of this book, however, we will use the phrase only for the most common type; the medium pieces (6, 8 or 9 pounders, and medium howitzers) which formed the backbone of the Divisional batteries. This will distinguish them from both horse and heavy artillery.

Horse artillery was composed of the lightest guns, 3 or 4 pounders (at least in theory; in practice they tended to become much heavier as the wars went on). All the horse guns had reinforced teams, and all their gunners were mounted. This gave great mobility to the batteries, so they were usually attached to cavalry formations. The exceptionally large number of horses, however, could create serious problems while the guns were in action: they were an encumbrance, as well as a good target for the enemy.

Heavy artillery, by contrast, consisted of the corps and army reserves; heavy howitzers, 12 and 18 pounder guns and even heavier guns in some armies, especially the Russian. Heavy artillery would typically be used to hold key points on the battlefield, for battering or counter-battery work, or to support a final mass attack.

Howitzers were short-barrelled, high-angle pieces which could fire explosive shells at long range, a grape-shot or canister at short range (the latter consisted of a number of small balls which spread out from the muzzle to give a huge shotgun effect). Guns, on the other hand, used direct fire with solid shot at either long or short range, or grape and canister at short range. The normal organization was in batteries of two howitzers plus four or six guns; with about a hundred gunners and drivers per battery, and a couple of dozen limbers, caissons, and other vehicles.

The Napoleonic skirmish

It was not very common in the Napoleonic Wars for small groups of men to run into each other in isolation. Because battles were generally fought in densely massed formations it was more usual for squads and sections to be packed together in the firing line along with the rest of their battalions or squadrons. If they came into contact with the enemy, it would normally be as part of a battalion or brigade battle, not in a skirmish of their own.

Small groups did sometimes meet, however, as in the following circumstances:

Scouting

One of the few ways in which generals could get information was to send out scouts; either small parties of light cavalry, or specialized staff officers (possibly with an escort). If one of these scouting groups ran into a similar enemy group or a static outpost, there might well be a skirmish. Thus a staff officer might try to reach a village post office to read the mail (a favourite means of gathering information), or to find a local guide. If he ran into a small enemy force, he might try to fight his way through.

Outpost action

Forces on campaign established security lines around their perimeters to warn them of enemy approaches, and to prevent infiltration. These lines could be manned by either infantry or cavalry, and would be

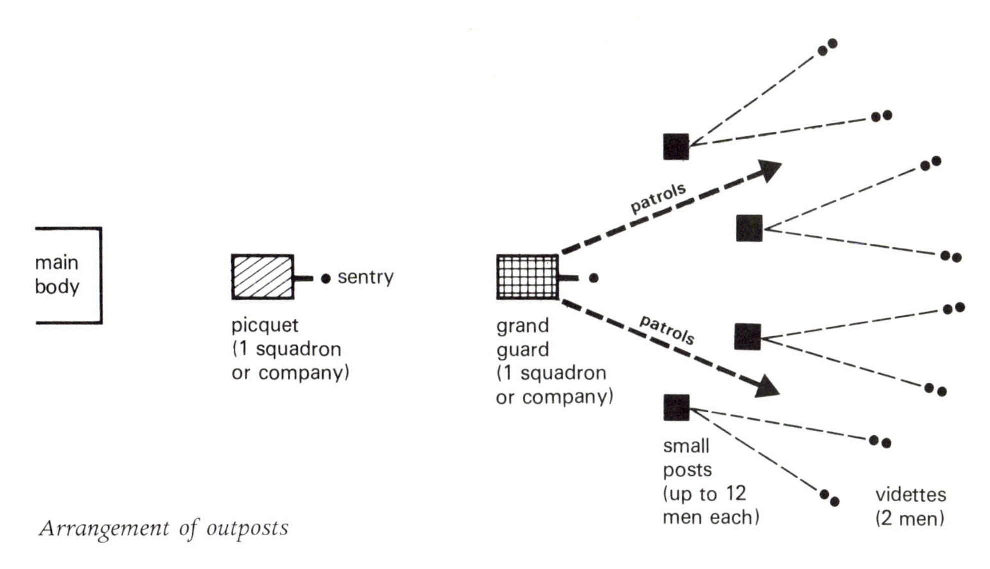

Arrangement of outposts

formed of four elements: two concentric rings of quite large posts, called piquets and grand guards respectively; and then a line of small posts (with perhaps a dozen men in each) which acted as centres for fans of videttes (each of a couple of men). This arrangement can provide endless variations for skirmish wargames, as infiltrators try to pass through, outflank, or otherwise surprise the videttes and small posts.

Foraging

Armies constantly had to send out small parties to collect food supplies from outlying farms and villages (either with or without payment). These foraging parties would usually take wagons with them and would be rather heavily laden. They would often need an escort, to protect them from enemy raiders or to persuade recalcitrant farmers.

Battles in difficult terrain

During battles in very close country, such as mountains, thick woods or heavily built-up areas, quite small segments of an army's skirmishing screen might easily become engrossed in separate little actions of their own and forget that they were part of a wider whole. In woods every thicket might need to be cleared; and in towns every room or staircase might be fortified, as they were at Saragossa (1809).

Guerrilla action

There were at least a few examples of guerrilla action in almost every Napoleonic campaign; either by small groups of soldiers, as in Prussia (1807), or by bands of armed civilians, as was especially true in Spain (1808–1813). Ambushes on couriers, supply convoys, or even small static posts, could lead to desperate skirmishes with no holds barred. Alternatively, an occupying army might launch expeditions to destroy known guerrilla camps in mountainous or heavily wooded areas.

What you will need to play the skirmish game

Now that we have glimpsed the types of troops, and the sort of actions they might fight, we can turn to the business of making it all into a game. The following items will be required.

A table top

This should measure at least 1m × 1.20m (3′ × 4′), and preferably more. A bed can be used at a pinch, although it may suffer from being a little too soft. The floor is also a possibility, but in this case players must take great care not to step on the models. As a general rule we can say that the better and more expensive the models, the less a floor game is advisable.

A means of representing relief

This can be based on one of four principles.

The classic military method is to use a sand table. This is very good for showing relief, but is laborious and messy to set up. It requires specialized equipment, and is not recommended.

You could make a cloth model. This consists of a cloth or thin linoleum sheet stretched over the table top, with relief shown by books, boxes, or blankets placed under the cloth. The visual effect can be very good, as hills swell naturally and continuously, and the cloth may have a realistically rough texture. It is often rather difficult, however, to persuade model soldiers to stand up on the hills, so you will need to be very careful.

If your table has a painted top, it may be possible to chalk a map directly onto it without damaging its surface. If this is the case the terrain can be shown very easily and quickly, and does not require any complex paraphenalia for hills. Some wargamers, however, object to chalked maps as unsightly and messy, despite their great convenience.

The most popular technique (but also the most stylized) is to use a table top with contours shown by flat cut-out layers of hill. Chipboard, expanded polystyrene tiles, thick cardboard and so on, can all be used to make the layers; and if they are carefully painted and detailed they can make very impressive scenic features. They have the great advantage that model soldiers can always stand up on each layer; but the fact remains that with this system you are left with a rather limited choice of relief. Unless you make a really large number of different hills, you will find that the same one keeps cropping up again and again in every battle. Another, less serious objection is that all your hills will have an artificially stepped appearance, so that it will require a certain amount of imagination to envisage the precise degree of slope.

A means of representing scenic features

Once you have established the relief of your battlefield, you will need to show woods, houses, roads and rivers upon it. Model railway shops stock a number of useful items for this, or you can make them yourself. Buildings can be cut out from card or built up from plaster or modelling clay. Woods can be made from dyed loofah, which is expensive but effective, or plastic sponge, which is nastier but cheaper. Rocks can be brought in straight from the garden; roads and rivers can be shown by suitably coloured tapes, or made from cork chippings, sawdust and so on. The precise methods used will depend upon the ingenuity and modelling skill of the individual, although particular care should always be taken to keep the scenery in scale with the model figures being used.

Two forces of model soldiers

These may be in either 25 mm or 54 mm scale (i.e., a man about 2 m (6 ft)

A model village, in this case being used in a Divisional game. Notice that each building, hedge and wall stands separately on its own small base. This allows a new configuration of scenic features be made for each game.

tall would be modelled by a figure either 25 mm or 54 mm tall). These are both popular commercial scales, although supposedly 25 mm figures may in practice turn out nearer 30 mm or 20 mm. For the sake of convenience, however, we will call all these sizes 25 mm. If the skirmish is to include a couple of dozen figures per side, then 25 mm will be the best scale; but if there is to be only a handful, then 54 mm may be preferred. The larger the scale the greater the detail on each model; but the greater, also, will be the cost of each model. The smaller scale also has the advantage that a given table top will represent a larger area of countryside, so we can use more ambitious flank manoeuvres, or more widely extended screens of men.

The models may be of metal or plastic if bought commercially, or of some rather more exotic material if the wargamer makes them himself. Cut-out paper soldiers have a long and respected past, while even prune stones have been used in some extreme cases. A fundamental problem with model soldiers is that the really good metal ones cost a great deal of money – perhaps twenty-five pence each – even before being painted. Plastic ones come cheaper, at around one or two pence each. With plastic, however, it is often difficult to find the precise type of Napoleonic figure that you are looking for; and even if you do, the paint has an annoying habit of peeling off muskets, plumes, etc. A wargamer with limited means would therefore be well advised to start with basic plastic armies, and gradually add specialist models in metal as his interest grows.

The two opposing forces should be planned to represent a realistic mixture of troop types for the sort of skirmish which is desired; and each man should, if possible, have a reference number painted unobtrusively on the upper side of his base. This will allow rapid identification in the course of the game and will facilitate the writing of orders.

Record sheets

Every player should have a record sheet on which he can write moves for each of his model soldiers. The sheet of paper should have the soldiers' names and/or reference numbers down the margin and their appropriate skill rating, with several lines between each for writing in abbreviated instructions for each move. There should also be some spare paper for writing any orders the soldiers themselves may issue.

Nuggets and rulers

When the decisions of chance or fate are called for in the game, we use a twenty-sided decimal dice to find the result. These may be called deci-dice for short; but I prefer the nickname nuggets, and shall use that throughout the following pages. These dice are available at many model or games shops and are especially handy for wargames. This is because they produce a score between 0 and 9, i.e., a choice of ten different scores. This makes it easy to generate percentages: if a 0 comes up, it stands for a percentage between 0% and 9%; 1 stands for a percentage between 10% and 19%; and so on up to 99%.

A ruler or tape measure marked in millimetres and inches is required for measuring ranges and moves. Retractable steel tape measures are probably the most convenient for this.

Playing the game

The learner should start with a small squad of perhaps half a dozen men under a corporal, and only attempt full section-size games when he has got the hang of the system of play. However many troops are used, there must first of all be agreement as to the exact type of skirmish to be played. The numbers and troop types must be decided; which of them represent the squad commanders; which have their muskets already loaded at the start of play; and especially, what objective each side is

Soldier	Ref.	Skill	Moves				
BERT	14	~~B~~ C	RUN	RUN	FIGHT	STAND	WALK
			FIRE	RELOAD(1)			
JAKE	15	~~C~~ ~~D~~ —	RUN	RUN	FIGHT	STAND	DIES
BILL (SGT.)	16	A	ORDERS	WALK	WALK	FIRE	RELOAD(1)
			RELOAD(2)	WALK			
HODGE	17	C	WALK	FIRE	RELOAD(1)	RELOAD(2)	RELOAD(3)
			FIRE	RELOAD(1)			
ADAM	18	D	WALK	FIRE	DIAGNOSE MISFIRE	CORRECT MISFIRE	FIRE
			RELOAD(1)	RELOAD(2)			

Example of a record sheet for the skirmish game, showing the first seven moves of a game. Bert has been wounded once on move 3, and so has Jake. Jake has been wounded again, mortally, on move 5. The rest have given fire support, but Adam had a misfire on move 2.

aiming to secure. Ideally there should be an umpire to settle all this (although he is not vital), and he may choose to reveal to each side only their own half of the picture, leaving enemy strength and position as a surprise.

At this stage we ought to know the ground scales being used. With 54 mm models, one inch represents one metre. With 25 mm models, one inch represents two metres. Each model soldier represents one real man; and each model building or scenic feature represents one real one.

The game progresses in a series of turns or bounds, each of which represents ten seconds of game time, even though it may actually take any length of real time to play through. In each turn the sequence of events is as follows:

(i) Players write down what each man is to do that turn.

(ii) Both sides simultaneously move their men, according to instructions. If the move includes any morale tests or the writing of any orders, that is done at this stage.

(iii) The result of any firing is calculated, including any firing in close combat.

(iv) The outcome of any close combats is calculated.

(v) Both sides agree that the turn is complete, and the next may begin.

Orders

At the start of the game, and whenever else command figures are supposed to issue orders (see movement rules, below), the player must make a written note of exactly what his figure has said to the other men. Whatever it is, their moves must thereafter conform to the orders. Only close combat, a wound or a new order may override an order once it has been issued.

Skill and wounds

Each man will be given a skill rating between 'A' (for a very expert fighter, a veteran, etc.) and 'D' (for a novice or untrained irregular). This rating will affect the combat efficiency of that man throughout the game, and will be reduced if he is wounded. For every minor wound his rating will be reduced by one grade. For a serious wound it will be reduced by three (i.e., an 'A' fighter would go down to 'D' if he suffered a serious wound, and a 'D' man would go down to 'G', etc.). After two serious wounds the man automatically dies.

Whenever any wound is sustained, the man falls over at once, and must test for morale. For a minor wound he spends the next whole turn standing up again, but for a serious wound he must remain on the ground, unconscious, for as many turns as are shown by one nugget roll. After that number of turns has elapsed he may stand up again in the subsequent turn. For example, Corporal Untel, skill rating 'B' (a moderately experienced soldier), suffers a serious wound after being hit by a crafty enemy rifleman. He immediately drops to a skill rating of 'E' and falls to the ground. He tests for morale by throwing a nugget (see below). Because he throws a 6 he remains in good heart, just. Had he thrown 5, he would have become 'scared' – and who would blame him? He must next find out how long he will remain unconscious on the ground. He throws a new nugget, and for a score of 4, finds he cannot start to stand up or move until after the fourth move from then. If the crafty rifleman happens to march up and bayonet him within the first four moves, there will be an automatic hit, and a new wound. If nothing of this sort intervenes, the corporal may stand up on the fifth move after he was originally wounded.

Movement

In each turn each man on foot may perform *one* of the following actions, and no more than one. These actions will be interrupted immediately if the man is wounded, engaged in close combat or fails a morale test during the turn (see combat and morale rules, below).

Walk	on level ground	12 m
	on rough ground	10 m
	in woods	6 m
Run	on level ground	18 m
	on rough ground	14 m
	in woods	8 m

All men who run have a nugget thrown for them. They trip and fall if it comes up 0 or 1. All men running are also automatically 'puffed' for one further turn for every turn they have run.

Turn about more than 90°; the first 90° is free.

Lie down, kneel down, or stand up, including standing up after tripping or being wounded.

Open or close a door.

Start to climb a hedge, fence, wall, etc., or to jump a small stream.

Finish crossing an obstacle, as above: i.e., the complete operation takes a total of two turns.

Issue an order verbally, or make a signal with drum, whistle, etc. The man issuing the order must be doing nothing else on that turn. The player must also make a written note of what was said in the order.

Aim and fire one shot with a musket, rifle, carbine, pistol, blunderbuss, etc. Loading is a separate process from aiming and firing and takes a total of three or, with a rifle, five turns. No shot may be fired unless the weapon has previously been loaded fully and in the correct sequence (see below). Even if this is so, all firing men must also have a nugget thrown for them. If it shows 0 or 1, there is a misfire.

Place cartridge in muzzle of weapon; the first part of loading.

move no.	musketeer 1 (no bayonet)	musketeer 2 (bayonet fixed)	rifleman
1	 successfully fire shot	 successfully fire shot	 successfully fire shot
2	 place cartridge in muzzle	 place cartridge in muzzle	 place cartridge in muzzle
3	 ram home cartridge	 ram home cartridge; throw dice because bayonet fixed	 ram cartridge
4	 prime and cock	 if not wounded by bayonet, prime and cock	 get out mallet
5	 fire, if dice for misfire allows it	 try to fire; fail if dice for misfire prevents it	 hammer home ram rod
6	 place cartridge in muzzle	 try to diagnose misfire	 fire, if dice for misfire allows it
7	 ram home cartridge	 correct misfire, if dice allows	 place cartridge in muzzle
8	 prime and cock	 fire, if dice for misfire allows it	 ram cartridge

Ram home cartridge; the second part of loading. If a bayonet is fixed, then the man must throw a nugget. For a 0 he suffers a minor wound to his forearm.

Prime and cock weapon; the third and final part of loading, except with the rifle, when it is the fifth and final part.

Get out ramming mallet for a rifle; the third part of loading a rifle.

Use mallet for ramming, with a rifle; the fourth part of loading a rifle.

Diagnose a misfire: if the weapon has misfired, there must always be a turn for diagnosis before taking further action with that weapon.

Correct a misfire: once a misfired weapon has been diagnosed, the fault must be corrected before the weapon may be fired. This process normally takes one turn, but a nugget must be rolled. If it comes up 0, 1, 2 or 3, then a second turn must be used, including a further nugget roll, which may prolong it still further.

Fix or un-fix a bayonet on rifle, musket or carbine.

Draw sword or any other weapon, e.g., a pistol, or replace it in scabbard/holster. This does not apply to cavalry carbines which are on handy slings.

Conduct one round of fencing with bayonet, sword, dagger, bare hands, etc. If two opposed players come within fencing range (2 m) all foot figures immediately stop whatever they are doing, and fight with whatever weapons are available. Thus in the first turn they may combine movement with either firing or a full round of fencing; but in later turns they must either move or fight.

Each cavalry figure may perform one of the following in each turn:

Start to mount or dismount.

Complete mounting or dismounting, the process taking two turns in all. When a cavalry figure is dismounted, he acts as if he were an infantryman.

Hold up to five horses; he may be either mounted or dismounted for this. If a horse-holder has to perform any other act, the horses will all bolt, and can be stopped only for a nugget roll of 8 or 9 for each horse, each turn.

Walk (mounted)	on level ground	14 m
	on rough ground	12 m
	in woods	6 m
Trot	on level ground	20 m
	on rough ground	16 m
Gallop	on level ground	28 m
	on rough ground	22 m,
	but rider falls off for 0 on one nugget throw.	

Jump a low obstacle; rider falls off for 0, 1, 2, or 3 on one nugget.

Opposite *The different loading sequences for the skirmish game*

Loading and firing weapons mounted is exactly the same as when on foot, although the horse must be stationary for all parts of the process.

Turn the horse through more than 90°; the first 90° is free.

Drawing swords or pistols, and fencing when mounted is conducted as for men on foot, except that the horse may continue to move while the man is fighting. If the horse is moving on a close combat turn it will not stop until the following turn (assuming the player wants it to stop). Cavalry duels will therefore rarely be prolonged affairs since horses will usually charge past each other in opposite directions. They will not often stand to fight on the spot. Horses may not be forced to charge over a man on foot: they will instinctively try to go round.

Orders may be issued when mounted if the man is not doing anything else at the same time. The horse may be either moving or stationary.

Morale

All figures within 10 m of an enemy (whether it is that enemy, or the figure itself which is actually moving) must immediately have a morale test to see if they become 'scared'. All soldiers who are wounded must also test for morale at once. To test morale, roll one nugget for each figure. That figure is considered 'scared' if the following scores, or less, appear: for skill rating 'A'–1; 'B'–2; 'C'–3; 'D'–4; 'E'–5; 'F'–6; 'G'–7.

Note that one grade is deducted from the skill rating if the soldier is 'puffed' or has suffered a minor wound; one is added if he is behind cover; and three are deducted if he has sustained a serious wound. Infantry attacked by cavalry must deduct two from their skill rating in the morale throw.

When soldiers are 'scared' they may not advance towards the enemy (i.e., they may not initiate close combat), and must stop where they are, at once. They may, however, defend themselves in combat, fire, or move in other directions on subsequent turns. They may not rally during the course of the game, since it represents too short a time.

Firing

When a weapon is fired, first make sure it has been loaded properly, according to the sequence already explained. Then roll a nugget for misfires (in rain there will always be a misfire). Next, find the range from the target, which may be long, effective, or close. Close range for all weapons is 2 m, i.e., the range of close combat. Other ranges are as follows (in metres):

RANGE	WEAPON				
	Musket	*Rifle*	*Blunderbuss*	*Carbine*	*Pistol*
Long range	200	400	40	100	40
Effective range	50	100	10	24	10

Now find if there has been a hit, according to the following table. A hit is scored if a new nugget roll equals or exceeds the score shown; for example, a puffed musketeer at 'B' status, firing at a static man in the open, would need a 7, 8 or 9 on the nugget to hit at effective range, but a 9 alone at long range.

FIRING MAN	SKILL	TARGET MAN Long range			Effective range		
		Static in open	*Moving in open*	*In cover or prone*	*Static in open*	*Moving in open*	*In cover or prone*
Composed	'A'/'B'	7	8	9	5	6	7
	'C'/'D'	8	9	–	6	7	8
	'E'/'F'	9	–	–	7	8	9
	'G'	–	–	–	8	9	–
'Puffed',	'A'/'B'	9	9	–	7	8	9
mounted,	'C'/'D'	9	–	–	8	9	–
or 'scared'	'E'/'F'	–	–	–	9	–	–
	'G'	–	–	–	–	–	–
Any two of:	'A'/'B'	9	–	–	8	9	–
'puffed',	'C'/'D'	9	–	–	9	–	–
mounted,	'E'/'F'/'G'	–	–	–	–	–	–
or 'scared'							

For all firing at close combat range, a hit is scored for a 5 or more (see combat rules, below).

Blunderbusses, at close and effective range, will always automatically hit every other man, however many are presented. If there is only one man, he will be hit. If two, one will be hit; if three, two will be hit, and so on.

Hits: hits may be scored by any firearm (see above), or in close combat with an *arme blanche* (see close combat rules, below). For every hit scored, throw one nugget for the effect on the target figure:

Nugget score		
	0–1	target stone cold dead
	2–5	target suffers serious wound
	6–9	target suffers minor wound

Close combat

Close combat will start when one or both sides move into close combat range (i.e., 2 m) of an enemy. During this turn, as the attacker (or attackers, if both sides advance simultaneously) moves into range, there will be *both* movement *and* close combat. In subsequent turns, however, there will be one or the other, but not both.

First find the result of any firing at close range, i.e., any participants in the close combat who want to fight by fire in this turn. Note that

A cuirassier attacks a Gordon Highlander in a skirmish game.

these participants may not also fight with other weapons in the same turn. Any fire within 2 m will score a hit for a 5 or more on one nugget, apart from blunderbuss fire, which automatically hits every other man within range.

Then settle the hand to hand combats, *i.e.*, the remaining participants in the close combat, after the firing troops have fired. The procedure is as follows:

Ascertain the skill rating for all non-firing men (after wounds), and adjust it according to the tactical circumstances:

Deduct one skill grade if the man is prone, facing away from combat, mounted, scared, or if the enemy is on higher ground, armoured or using a longer weapon: Cossack lances are the longest, then ordinary lances, billhooks, swords, bayonets, daggers and bare hands, in that order.

Deduct two skill grades for a man on foot against a mounted man walking.

Deduct three grades against a mounted man trotting or galloping.

Each man in the fight will now have a total score, representing his personal ability to strike at the enemy. We now find whether his blow has been effective, by rolling a nugget for him. He inflicts one hit on the enemy if his score equals or exceeds the following: skill 'A' – 4; 'B' – 5; 'C' – 6; 'D' – 7; 'E' – 8; 'F' – 9; 'G' – no chance. If the target figure is unconscious, a hit will be automatic, and no dice need be rolled.

Now find the damage inflicted by each hit, exactly as for hits from fire (see above). The combat may continue for as many turns as there are combatants within striking range of each other. Repeat the above process once for every man each turn.

It will be found that there are two alternatives in the skirmish game. Players may want to rush in madly against the enemy, suffer his fire, and stake all upon a desperate close combat. On the other hand, they may choose to approach cautiously, using cover and engaging the enemy in a fire fight. Depending on the circumstances, either approach may be preferable: the precise choice will depend upon the balance of forces, the terrain and – naturally – upon the skill of the player.

3
The Divisional game

Our next game is a great deal more technical than the skirmish, and is in fact the most complex in this book. It is the Divisional level wargame, where the player represents a Divisional commander with control over a number of brigades. He might have between 2,000 and 12,000 infantry under his command, which at a ratio of one model soldier to thirty-three men, works out at between 60 and 360 model soldiers. There might also be from one to four batteries of artillery; a cavalry regiment or two; and the necessary engineer, medical, veterinary, military police, and headquarters support. In a cavalry Division there might be between one and four thousand cavalry, with a horse battery or two, as well as the various headquarters elements.

Despite its relatively high level of complexity, or perhaps because of it, the Division is the most popular subject for Napoleonic wargaming, and there are already many sets of rules for playing it. The models are normally 25 mm or 15 mm scale, which will allow a Divisional battle to fit conveniently on the average dining-room table. In some cases the game may also grow to quite huge proportions, since it requires only 1,000 model figures to make up a big army corps of 33,000 men; and with a few army corps you have an army. The biggest game of this type I have heard about involved no less than 6,000 model figures (representing 200,000 men) on each side, deployed over the floor of a community centre. Battles on this scale are very much the exception, however: major events to be planned many weeks in advance, and involving dozens of players. For normal use it is the Divisional battle which seems to have swept the board, since at this scale the game can be played informally in a single evening.

The Napoleonic Divisional battle

In most Napoleonic battles Divisions fought as parts of armies and army corps, rather than entirely on their own. It is true that there were a few cases of truly Divisional battles being fought in isolation – e.g., at Maida in southern Italy (1806) – but for the most part single Divisions tended to come into action only in vanguard combats, when at least one of the two sides could bring up significant reinforcements during the course of the battle.

When both sides had reinforcements at hand, the initial clash would often be the signal for all other forces in the area to 'march to the sound of the guns'. The battle would thus steadily grow in size, and be taken out of the hands of the original Divisional commander. The decisive battle of Auerstadt (1806), for example, began in this way. Gudin's Division of Davout's corps arrived in front of the enemy, started fighting, and was then gradually reinforced by the remaining Divisions of the corps. In a sense, we could say that Gudin's Divisional battle lasted only as long as he remained the senior commander present, since his Division soon became submerged in a much bigger fight. If we were to make Auerstadt into a Divisional wargame, therefore, we should presumably have to stop the game as soon as Gudin was reinforced.

On many other occasions only one of the two sides was in a position to reinforce its original Division. As time went on, this side would remorselessly bring in more and more fresh formations, while its enemy was being equally remorselessly worn down. However well or badly it fought, the smaller force would eventually be forced to retire to a new position in the rear, and then possibly to another and another after that. The best that its commander could hope for was to impose as much delay upon the enemy as possible. Both Bagration at Schöngrabern (1805) and Delaborde at Rolica (1808) fought extremely well; but both were doomed from the start to being forced back by superior numbers.

It was typical of Divisional battles that one side was hopelessly outnumbered by the other, and we should try to reflect this in our choice of wargame scenarios. The issue was not, which side would hold the battlefield at the end of the day, but which commander could put up the better show with the forces available to him.

The smaller force might have time to strengthen its resistance with field fortifications, as several of the French Divisions did in the Pyrenees in 1813. This would impose a more careful approach on the attacker, leading to a more formal style of battle. On the other hand, the defender sometimes saw that his only possible hope of survival was a fighting retreat at full speed. Pacthod's 4,000 infantry at Fère-Champenoise (1814), for example, fought for hours on end to shake off an overwhelming cavalry pursuit, and some of them succeeded. If one had enough playing space available, this would surely make a fascinating wargame.

During the battle itself, both sides would normally deploy behind a screen of infantry and cavalry skirmishers. These would probe and worry the enemy defences, seek out the flanks and report information back to the commander. Under their cover the Division's main body would be drawn up.

Napoleonic Divisions normally deployed on two or three lines of battalions, with either a separate brigade in each line, or some parts of

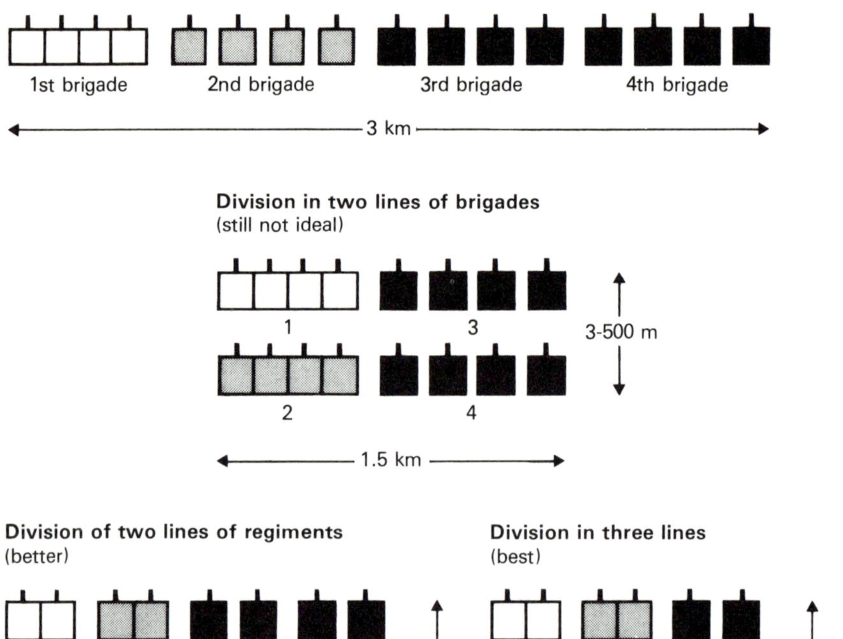

Division formations; each symbol represents a battalion

each brigade. Opinions varied on the best method; but it had often been found from experience that a single line of battalions was a very weak formation, and required some sort of reserve. Putting one brigade behind another was damaging to morale, since the men of the front line brigade would be unfamiliar with their supporting line, and would feel uneasy. Also supporting the front line would be the massed artillery of the Division, sited at the most commanding spot, and perhaps some cavalry positioned to the rear.

The formations used by each battalion would vary enormously, and each battalion was an entity complete in itself, with only a moderate interest in what was happening on its flanks. In theory the battalion line three deep tended to be favoured for the defensive, while the

Infantry at the point of contact, or rather non-contact. This picture shows very clearly that a bayonet charge gained its effect from neither firepower nor actual hand-to-hand fighting, but rather from 'the moral power of steadiness . . . over disorder which stupifies itself with noise'.

battalion column of divisions was preferred for the attack. To both of these rules, unfortunately, there were innumerable exceptions, and it would take a whole book to explain them all in detail. It is a curious fact, by the way, that no book devoted entirely to Napoleonic infantry tactics has been published since 1902. Perhaps all that need concern us here is that, firstly, since it was particularly vital to cover a battalion's flanks against cavalry attacks, a solid column or square was often used for defence whenever cavalry seemed to be threatening; although this was an especially weak formation against infantry or artillery.

Secondly, many experiments were made in the use of columns heavier than a single battalion; but this almost always led to failure. The additional numbers in regimental or brigade columns added nothing to the impetus of the attack, but served only to make a more unwieldy formation and a better target.

After more or less softening-up by artillery and skirmish fire, the attacking forces would send a first wave of battalions to assault the enemy. Everything then depended upon how these troops reacted when they came within musket range. If they pressed on with determination, the defenders would probably shoot badly and start to melt away. If the defenders stood firm, on the other hand, then the attackers might well grind to a halt and try to use musketry fire instead of their forward impetus. One side or the other would lose its confidence before anyone actually came into hand-to-hand combat,

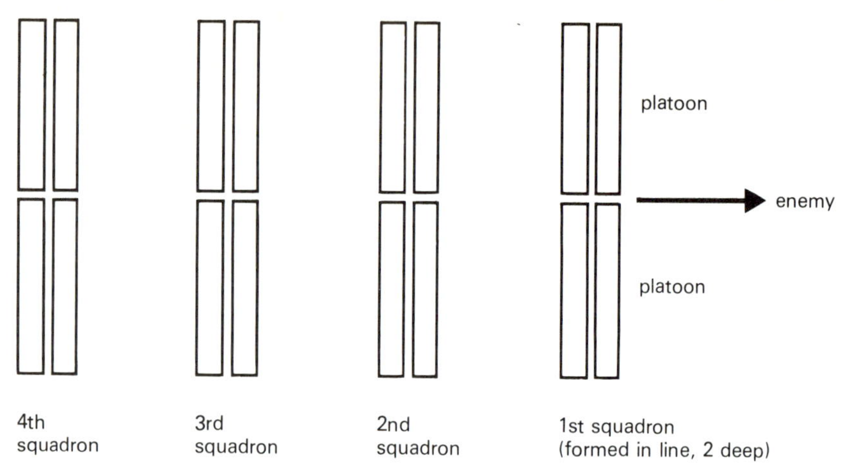

Typical formation of a cavalry regiment

and it is a well-documented fact that there was very little true bayonet fighting during the Napoleonic Wars. The expression 'bayonet attack' is rather misleading: it tended to stand for 'a menacing forward movement without much firing', rather than a genuine expectation of actually skewering anyone.

If the first assault cleared away the enemy, well and good: the cavalry could pursue. If the first attack faltered, on the other hand, there would probably be a long and indecisive fire fight between the units directly concerned. The victory would go to whichever side could feed in fresh troops to break the deadlock. Thus at Talavera (1809) the first French wave was repulsed by the British, but the British were then in turn thrown into confusion by the arrival of the second French line. At Albuera (1811) the French first wave again faltered; but this time it was British reinforcements who arrived first, to repulse the French. A very great deal depended upon who had the freshest reserves at hand, and how well they could be fed into the decisive combat.

Cavalry action tended to follow the same lines as infantry fighting, although faster movement and less emphasis on musketry led to more frequent clashes with cold steel. Whether or not there were such clashes, it was once again the timely arrival of formed reserves which proved decisive almost invariably. Keeping some 'sabres in hand' was the only sure way of winning the cavalry fight, so most cavalry regiments would attack with one or two of their squadrons held back from the front line. Against infantry, on the other hand, horsemen found that their mounts were reluctant to charge home into formed units. Most would shy away at the last moment, and it was only against unformed or broken infantry that they posed a serious threat.

As for the artillery, it would usually be massed, firing happily away at the largest targets which were presented, and remaining better under control than almost any other section. In the attack its usefulness would often end soon after the assault had been launched, since the attacking troops would mask the target from the guns. Horse artillery was sometimes used to accompany assaults, but this left it very vulnerable to the defensive fire. In the defensive, artillery could lay down a devastating curtain of fire at both long and short range, provided the gunners had not been scared away by the enemy's advance.

What you will need to play the Divisional game

A table top, a means of representing relief, and scenic features will be needed, exactly as for the skirmish game.

Two model Divisions

These should be in 25 mm or 15 mm scale. As with the skirmish game, the smaller the scale, the larger the battle which can be fitted into a given area; but less individual detail will show on each figure. For a Divisional game it is best not to mix nationalities too much, since most nations were usually able to field quite large independent formations. The British, however, are a striking exception to this, since their formations in the Peninsula regularly included Portuguese troops, along with some Spanish and Germans. In the Waterloo campaign there were Dutch-Belgians, instead.

The wargamer should try to maintain reasonably realistic proportions between the different types of troops, and should especially resist the temptation to include isolated battalions of Guards within line brigades. Guards usually operated in massed formations of their own, and would seldom be used in penny packets alongside the much scruffier line or militia units.

When we work on the basis of one model to thirty-three men, we find that battalions can be between ten and thirty models each. If desired, a different strength may be used for each battalion, although for convenience it is useful to have a standard battalion strength of twenty. Cavalry regiments work out at about four squadrons, each of three figures; while artillery is shown by one model cannon plus four model gunners per battery, representing six or eight pieces, and just over a hundred gunners and drivers.

Figure bases and formations

Once we have collected our models, it is best to mount them on bases, to make them easier to manoeuvre in large numbers. The bases can be of stiff card, plastic card, or even tin sheeting. The soldiers should be stuck on with a strong epoxy adhesive, and then the bases painted to represent green grass or brown earth. An unobtrusive number should be painted on the upper side of each base, for ease of identification.

Infantry should be divided into groups of five models, i.e., 165 men. According to the particular national organization of that army, each of these groups will represent either one company or two, but for the sake of simplicity we will hereafter assume two companies, i.e., a division. The models for each division are stuck onto a base in a single line. The base should be as near to 50 mm long as is possible, and as thin as the base of the figures will allow. This represents the division in line three deep, with officers and N.C.O.s acting as a fourth line in the rear. British infantry bases should be longer, say 70 mm, to represent the British formation two deep, with officers and N.C.O.s forming a third line.

There is a slight problem posed by the different sizes of 25 mm and 15 mm figures. In theory we should use smaller bases for 15 mm models, but for convenience it is better to use the same size bases for both; allowing the 15 mm troops plenty of elbow room, but slightly squashing the 25 mm soldiers. As the figure scale is bigger than the ground scale, this does not really matter (see discussion of scales).

In the course of the battle, the four division bases in each battalion may be arranged in certain combinations to represent the various formations which are possible in battalion drill:

Battalion line is represented by all the divisions in the battalion being laid side by side. It is a good formation for fire, but not for movement. If another friendly unit stands within 50 m behind the line, they will be considered to have become entangled together, and 'bunched'.

Column of attack is represented by either two divisions abreast, two deep, or a one-division front, four deep. This formation is handy for movement, but vulnerable to fire. It also becomes 'disorganized' if it 'bunches' too close to another column on its flank, less than 50 m away. The same is true of units which follow a column at less than 50 m distance; they will be considered entangled and 'bunched' with it.

Column of route is the only possible formation for road movement or crossing bridges. The four divisions of the battalion are laid end to end so that they appear to be going sideways along the road. No more than one figure abreast will be allowed on roads or bridges.

To form square, the four division bases are arranged in a hollow square.

If a battalion sends forward one division (normally the flank companies) to skirmish, that division may be retained under control as long as it advances no more than 400 m ahead of the main body. If it goes further afield it will be lost to the main body, will become 'disorganized', and may not rejoin unless it throws a rallying dice score of 9. More than a single division may be advanced, if desired, under the same rules. An interval of at least one base-length (50 m) should be left between each skirmishing division to represent the wide dispersion of

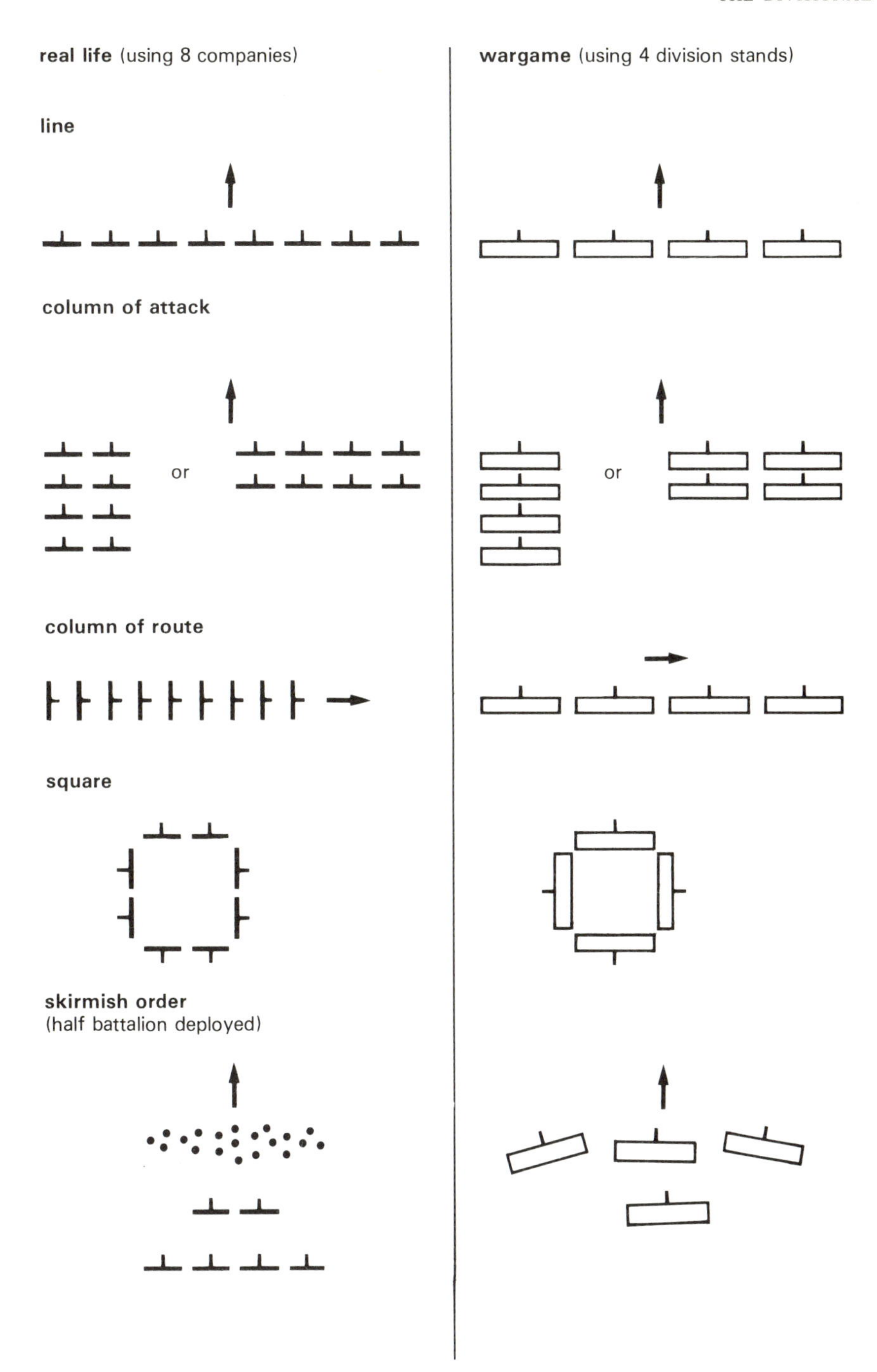

Battalion formations

This scene illustrates the various groupings of figures on bases for the Divisional game. On the road there is a limbered gun team on the march, then a battalion in line or column of route (four bases each of five men), then a wagon on a base. In the middle rank is a battalion in square, then a battalion in close column of divisions, then a battery in the firing position (four men and a gun on a base), with a courier and a two-man staff base behind, and then a cavalry regiment of three squadron bases, each of three men (note the formation is two-up). In the front line is a battalion of deployed skirmishers.

the skirmishers. Any troops may skirmish, and in any numbers. Whole Divisions skirmishing were not unknown in the Napoleonic Wars; the only distinction was that poor troops skirmished badly, while good troops skirmished well.

Each cavalry regiment is divided into a number of squadrons, each of three models, totalling ninety-nine men. Each squadron has a base about 50 mm long, and as thin as the bottom of the figures will allow. The figures, as with the infantry, are arranged in a single line; although in this case it represents a squadron in double line, the normal fighting formation.

Each regiment should act as a whole in most circumstances, and should not detach individual squadrons. The only exception to this is when squadrons are detached for skirmishing or scouting. For massed action in battle they would return to their regiments.

A decision must be made for each regiment whether it wants one, two, three or four squadrons in the front line (or more, if they are available). Whichever it chooses, these will receive the first shock of battle, while the remainder act as a reserve in the second line.

For movement on roads cavalry adopt the column of route; i.e., each squadron turns sideways on, to follow the road one abreast.

Each model gun is placed on a base 50 mm wide and as thin as the model will allow. Four model gunners are ranged around the gun on the base. This represents a battery in line, facing the direction in which the model gun is pointing. It may be either unlimbered, when the gun is pointing at the target, or limbered, when the gun trail is pointing along the line of movement. Players should always specify clearly which is being used at any particular time. Batteries may move along roads, when limbered, without any further contraction of frontage.

Couriers are represented by single models of staff officers or cavalrymen. Commanders and their headquarters are represented by two-man groups on a single base, apart from the Divisional commander, who has a four-man base. It is essential that every player must know exactly which figure represents him in the game, so that the location of his HQ may always be known.

Each model wagon, plus one or two drivers, is placed upon as small a base as possible. This represents a company of thirty wagons.

Stationery

As in the skirmish game, a set of nuggets will be needed, and a ruler marked in millimetres and inches. Pencil and paper will also be required, both for writing orders and for recording the changing status of units. Some wargamers are reluctant to allow paperwork into their Divisional games at all, because they confuse paperwork with complexity. It is quite true that games may become too complex if the paperwork gets out of hand; but it is also true that a reasonable level of written records can greatly simplify the mechanisms of play. We will accept the use of written records if they are kept to a minimum.

Playing the game

A scenario is devised whereby the two opposing Divisions have some realistic reason to join combat on the model battlefield. The battlefield is then laid out, including all model soldiers who start the game in open positions, not concealed from the enemy. Troops starting in covered positions are not laid out, but their positions are noted on paper.

A good way for experienced players to lay on a battle scenario is to organize a map campaign (see the chapter on free kriegspiel). This does not have to be a big or elaborate set-up, and some very good battles can result from only an hour or two's manoeuvring on a map, provided that everything is kept simple. Normally, however, the Divisional game is best suited to a contrived, one-off scenario, especially designed for that particular occasion. The scenario may be taken from a true historical combat or, more usually, it may be invented by the wargamer as a possible or typical confrontation, which might have happened. If it is invented in this way, however, beware of making it too even. There is nothing less realistic than combats exactly equally balanced between exactly equal forces, on exactly equal terrain for both sides, with exactly equal orders and aims. That situation never arose in real life.

Scenarios may start with both sides already in close combat, or they may require both forces to march onto the battlefield little by little, preceded by scouts and advanced guards. In the former case heavy fighting will develop from the very start of the game, and the action will be fast and furious. In the latter case the game will be longer, and will build up more gradually. To a large extent, therefore, this choice depends upon how much time is available for the game.

As far as possible, at the start players should be kept in ignorance of the enemy's strength and aims. Each player will then have to use his intelligence, in both senses of that word, to find out what his opponent is up to.

Orders

When each Divisional commander has been given as much (or as little) knowledge of the scenario as he needs at the start of the game, he writes general orders for all his subordinate commanders. Nothing may be done by these subordinates in the course of the game unless it can be shown to be consistent with the original orders, or with subsequent orders arriving from the Divisional commander, e.g., by courier.

Scales

There are four types of scales which are essential if we are to understand the relationship between time and distance in this game. These are the ground scale, the figure scale, the vertical scale and the time scale.

The ground scale is used for all ranges, movement rates, unit frontages and other horizontal measurements. To fit a reasonably-sized action on to the average table, it is convenient to say that 1 mm on the model represents 1 m on the real ground. Thus a cannon with a range of 1,000 m would be allowed to shoot up to 1,000 mm (about 3 ft 3 in) in this game.

The figure scale follows from this, because the units we have arranged on division and squadron bases will now have their frontages represented by the frontages of the bases. If the base has a 50 mm frontage, that represents a unit 50 m long. This was in fact about the frontage of a division in the Napoleonic Wars; so we can say that each five figures must represent a division (165 men), i.e., each figure represents thirty-three men. Using the same methods we find that each model gun represents a battery, and each model wagon represents a company of wagons. The only exception is for couriers, who act as individuals. In their case we stretch the scales a little, and say that each model courier represents one real one.

We find that each model house 60 mm × 60 mm will represent a group of buildings (perhaps a hamlet or large farm) with an area of 60 m × 60 m. A group of model buildings may represent a whole small town. So far so good. The only trouble comes with the vertical scale, since we now find that the model soldiers and buildings are towering over the battlefield like so many Colossi and Empire State Buildings.

The vertical scale of the models is apparently 15 mm or 25 mm; i.e., a man about 2 m (6 ft) high is represented by 15 or 25 mm. This works out as $7\frac{1}{2}$ mm or $12\frac{1}{2}$ mm equalling 1 m, which is of course very different from the ground scale being used. Unfortunately there is nothing much we can do about this discrepancy. It is just a penalty we have to pay if we are to use detailed models for this sort of game, so if we ever have to measure vertical distances we must keep it in mind. When units want to see over the heads of others, for example, we should calculate their height according to the ground scale; not according to the vertical scale. So, a man will theoretically stand only 2 mm high in the model, rather than the 15 or 25 mm which he actually stands. Buildings will theoretically be 10 mm high rather than 75 or 125 mm, and so on. Only in the case of terrain relief itself will we take the contours of the model battlefield as accurate to the ground scale. If a hill is 20 mm high on the model, therefore, it represents a height of 20 m.

As for the time scale, one turn in the game represents two minutes in the battle, even though it may take any length of real time to play.

Status, losses, and basic morale

These things form the guts of the rules, and must be followed carefully if the rest is to make sense. Each unit is initially allocated a basic morale, a state of training, and a certain number of effective men. To represent all these things together, for simplicity, we allocate a single 'status' grade to each unit at the start of the game.

Guard or élite units are grade 'A' status. If they have to detach a sub-unit – a division detached from a battalion, or a squadron detached from a cavalry regiment – that sub-unit will be status 'D', but the status of the parent unit will remain unchanged. Such detachments should normally be avoided, except for specific skirmishing tasks. Line troops are 'B' status, their sub-units are 'D'; militia units are 'C', their sub-units are 'E'; irregulars start at 'D', their sub-units are 'E'.

As the battle progresses, units may lose status due to fatigue, failing various morale tests, or because they lose casualties to enemy fire. Every loss of one status grade represents the loss of about ninety infantry casualties to all causes; forty-five cavalry casualties; or one gun and its crew from a battery. These men are not all necessarily killed or wounded, however, since a very large proportion of them may have fled from their unit, physically unharmed. Napoleonic battlefields were notorious for the crowds of stragglers lurking in the rear, well away from the fighting line.

In these rules, units may lose a number of tenths of a status grade, as a result of certain forms of enemy action, e.g., artillery or skirmish fire. In this case the loss is recorded, but it becomes important only when losses of ten tenths have been accumulated by a given unit. For example, a 'B' grade unit which loses 0.7 of a grade will still fight as 'B' grade. If it then loses 0.3 of a grade, it will fight at 'C' grade.

The status levels represent the following attitudes:

Status 'A' Completely confident unit.
'B' Slightly apprehensive unit.
'C' The unit is somewhat worried and/or damaged.
'D' The unit is badly alarmed and/or hurt but still in the fight.
'E' The unit is now rated as permanently 'shaken' (see morale rules, below). It may not advance towards the enemy, although it may perform other manoeuvres. If it is within range of the enemy, it will fire wildly until it is moved out of range.
'F' The unit is 'broken' and must retreat as fast as possible to a safe place, in a 'disorganized' skirmish formation.
'G' The unit has had the stuffing knocked out of it completely. Only a small nucleus of brave men remains with the colours, of no tactical significance. The model unit is removed from the table.

Note that no unit is removed from the table unless it falls to status 'G'. There will thus be no visible means of telling the status of an enemy's unit, apart from the way it behaves. This is only realistic, since in battle it is always difficult to tell the enemy's precise status.

Unit	Original status	Current status
200 INF.	~~B~~	~~-0.5~~ C -0.2
201 LT INF.	~~A~~	~~B~~ C ~~S~~
202 GUARD INF.	~~A~~	B -0.2 S
203 INF.	~~C~~	~~D~~ ~~E~~ ~~F~~ G removed ~~S~~ ~~S~~ S
204 INF.	C	-0.2
205 LT CAV.	B	
206 HEAVY CAV.	B	
207 FIELD BTY	B	~~-0.8~~ -0.9

Example of a record sheet for the Divisional game.

The 200th infantry battalion has lost 0.5 of a status grade, then a further 0.7, bringing it from 'B' status to 'C' −0.2.

The 201 light infantry has lost one status grade and then another, later. At some stage it was 'shaken', but it has now rallied.

The 202 guard infantry battalion has lost 1.2 grades, and is still shaken; whereas the 203rd battalion has had no luck whatsoever.

The remaining units have suffered light losses, but have not yet lost any full status grades.

To keep track of your own losses, a written record is kept. If there is an umpire this task is given to him, and he might not always choose to release the information about status, even to friendly commanders. Lined paper is used for this record keeping, with a list of all units down the left-hand margin. Inside the margin is the unit status at the start of the game. As units suffer losses the new status is recorded along the line, and the old one crossed out. When a unit becomes temporarily 'shaken' due to a morale test, the letter 'S' is added at the end of the line, and crossed out if the unit is successfully rallied.

It does not take much practice to keep these records quickly and efficiently. They give an instant summary of the state of the Division, and at the end of the battle they can be used to add up the casualties suffered.

Temporary morale changes

The status of each unit gives an accurate reading for basic morale and losses. There are certain circumstances, however, when the basic score is temporarily changed due to a passing influence. In these cases the unit may become temporarily 'disorganized' or temporarily 'shaken'.

Any unit which is in the process of changing formation, lying prone, 'bunching', crossing an obstacle or skirmishing too far from its parent unit, will count as temporarily 'disorganized' until it stops that activity. This means that the troops will still obey orders, but as a result of the particular operation they are performing, they will lack their full steadiness. If the enemy attacks while they are 'disorganized', they will fight as if they are 'shaken' and may throw a dice for a quick rally. If they are not attacked they will be reorganized automatically as soon as their particular operation is complete.

Units which count as 'shaken' are those which are not fully organized for combat. They may not advance towards the enemy, although they may still carry out other manoeuvres. If they are within range of the enemy they will automatically fire at him until they are moved out of range (for the effect of this sort of fire, see below).

A unit will always be 'shaken' if its status has fallen to 'E' or below. In this case it is 'shaken' permanently, and may not be rallied. A unit with a higher basic status may also become 'shaken', however; but only temporarily, as it may later be rallied. Units become temporarily 'shaken' in this way either as a result of combat, or after a morale test.

Any unit which loses a close combat is automatically 'shaken', whether or not it also loses status. As long as it stays above 'E' status, rallying may be attempted.

Morale tests are taken in the following circumstances:

When the unit is within musket range (200 m) of an enemy unit or sub-unit.

A friendly unit or sub-unit attempts to pass through the unit; this does not include the unit's own skirmishers returning to re-form with the unit.

A friendly 'broken' unit, i.e., at status 'F', moves past the unit.

Enemy infantry achieve a rear attack on the unit, or enemy cavalry achieve a flank or rear attack. In neither case will a morale test be required if the defending unit had previously faced a sub-unit towards the attack, e.g., cavalry cannot claim a flank or rear attack against a square.

The Division commander is killed or wounded within 200 m of the unit.

Cavalry which have just won a close combat must test morale, as they may become over-excited and liable to pursue the enemy recklessly.

The morale test itself consists of rolling one nugget. The unit will lose the morale test, and become 'shaken', for the following scores, or less:

Unit's status	*Nugget score*
'A'	1
'B'	2
'C'	3
'D'	4

Rallying When a unit is temporarily 'shaken' or 'disorganized', for whatever reason, it must fight as if it were at status 'E', unless it can achieve a quick rally. In this case a nugget is thrown before the result of combat is calculated. If the score is 9, the unit rallies quickly, and fights at its true basic status level.

'Shaken' units are normally rallied more slowly than this. One nugget is rolled for each temporarily 'shaken' unit on each turn. Any unit which scores a 9 may be rallied, starting at the beginning of the next turn.

Linked to morale are the initiatives which unit commanders may wish to take in a crisis, contrary to any previous orders. These initiatives may occur during close combats, when the units have been caught in the wrong formation or with the wrong orders, e.g., infantry may not be in square against cavalry, or may have orders to stand still when it really wants to counter-attack, etc. In these cases the player must wait until half-way through the turn before he may change the unit's orders at all. Then he may claim the 'combat initiative' to perform the new act. The initiative will be allowed only if the following score is equalled or exceeded in one nugget roll:

Unit's status	*Nugget score*
'A'	5
'B'	6
'C'	7
'D'	8

If these scores are not achieved, the unit must follow the original orders. If the scores are achieved, the unit may start to carry out the new initiative during the second half of its turn.

Sequence of actions in each turn

(*i*) All players decide what they intend to do during the turn. If the opposition is especially cantankerous it will be necessary to use markers or make a written note, but this is not so with sensible players.

(*ii*) All players simultaneously move all units they had decided to move. Every unit may be moved in each turn, unless it is to fire on that turn, or is unavoidably detained by the morale rules.

Because movement is simultaneous, players are on their honour to carry out the moves they had originally planned, and not to change them. This often causes some problems with ungentlemanly players who try to take advantage of what they see the other side is up to. Several methods may be used to limit this form of cheating, however, and in extreme cases recorded moves may be adopted. The use of an umpire is a much less cumbersome solution, and so is a rule that every player must start by moving the pieces on his right, and then work across to his left. In this way each player will always move half his pieces before the enemy's opposite number, and half afterwards, thus maintaining a rough balance. Another useful rule is that once a player has moved a piece, he must not change his mind and move it in some other way.

(*iii*) Both sides announce which units are firing, and at what targets. All firing units are assessed on the basis of their status at the start of the turn, so that any damage inflicted upon them during that turn affects their firepower only on the following turn. Artillery fire, however, should as far as possible be assessed before musketry.

(*iv*) Both sides take any morale tests which are due, and settle any close combats, including crisis initiative moves.

(*v*) Both sides attempt to rally any temporarily 'shaken' units.

(*vi*) Players then, and only then, take receipt of any orders arriving during the turn, and write any new orders to be dispatched by courier during the next turn.

(*vii*) When all this has been finished, someone with a loud voice (or the umpire if there is one) shouts that the turn has finished and the next one is starting. Players will keep a written record of which turn is in progress.

Movement

Units move the following number of metres during each turn:

Type of unit	*Open ground*	*Rough ground*	*Obstacles i.e. temporarily disorganized units*
Formed infantry in line	130	90	60
Formed infantry in column	170	110	40
Formed infantry in square	30	10	—
Skirmishers or irregulars	200	140	80
Field arty or wagons	100	40	—
Horse arty or heavy cavalry	250	200	50
Light cavalry or staff group	300	250	100
Couriers	350	300	200

Concealed moves: any move which is ordered in a part of the battlefield that should be invisible to the enemy, e.g., behind a hill, or in a village or wood, will not be laid out on the table top. The figures concerned will be kept off the table, until the enemy has some troops in a position to see them. While troops are moving (or standing) concealed in this way, their positions must be reported to the umpire (if there is one) or written on a piece of paper. Any troops which fire from a concealed position must immediately be placed on the table top, since the smoke of their firing would automatically betray their positions.

Roads: all units add 10 m to movement rates on roads, but must be formed in single file or column of route.

Slopes: count as rough ground, whether moving up or down.

Heavy ploughland: counts as rough ground.

Open woods: count as rough ground.

Thick woods: count as obstacles. Troops automatically form into skirmish order if they enter either type of wood. Regular formations are impossible in this type of terrain.

Marshes: count as obstacles.

Rivulets: any unit trying to cross in that turn must move at rough ground rate throughout the turn.

Streams: count as obstacles throughout the turn.

Small rivers: may be crossed by infantry and cavalry units in four turns, but are totally impassable to guns or vehicles, without a bridge or ford.

Big rivers: totally impassable to all except couriers, who may swim across in four turns, but are drowned if they throw 0 on a nugget.

Bridges and fords: all units cross these, provided they are formed in column of route.

Villages: all movement through villages counts as movement in rough ground, whether or not there are roads through the village. If troops wish to enter or leave houses, it takes a whole turn in each case. Small houses will accommodate the equivalent of half a battalion; large houses take a whole battalion. Precise capacities for each building should be agreed before the start of the game.

Low hedges, walls, ditches, etc., count as obstacles to infantry, but are totally impassable to cavalry and artillery, unless engineers have opened a breach.

High walls are impassable to all units.

Mounting and dismounting: cavalry take a whole turn to mount or dismount.

Limbering and unlimbering: artillery units take a whole turn to limber or unlimber their guns, during which the guns may not be fired. In this game we do not use model limbers, so players must make specific announcements as to whether their guns are limbered or not.

Manhandling and traversing artillery: artillery may be moved by hand, without first limbering up, at the rate of 30 m per turn. Alternatively it may be traversed through an angle of 45° in half a turn, to face a new target. Any battery firing upon a target must face within 45° of it.

Infantry lying down: it takes one whole turn to make an infantry unit lie down, and another whole turn to make it stand up. During both of these moves the unit is temporarily 'disorganized'.

Changing formation: units which are to change formation move each of their sub-units to their new positions in the normal way, taking the time it would normally take to move troops that distance. No troops in the unit may perform any other act, e.g., firing, while any of the unit's members are still in the process of changing formation. The whole unit is also temporarily 'disorganized' throughout the operation.

Firing: units, including cavalry, may either fire in a turn, or move. The only exceptions to this are:

Artillery which is traversed or manhandled for half a turn may then fire for the remaining half turn, but only at half its normal effectiveness.

Units which have elected to fire in a turn, but which receive an enemy attack during that turn, do not fire in the normal way. Instead, the effect of their fire is included within the same calculation as the effect of the close combat itself.

Rifle-armed units fire only once every two turns. Between each shot they must take one stationary reloading turn. Pedants may argue that Napoleonic troops could fire much faster than we allow in our rules. It was true that this was possible with very well drilled troops, in peacetime, and without counting the delays imposed by command and control problems. In battle, however, with average troops, the rate of fire must have been much nearer what we suggest here.

Couriers: every time a courier is sent with a message, he receives it at the end of one turn, but does not start his actual movement until the following turn. A nugget is then thrown. If the score is 1 or 2, he moves at half rate for the whole of that mission (each courier has only one nugget rolled for him per mission). If the nugget score is 0 he is killed, injured, or hopelessly lost on the way, and never delivers his message. If a courier does arrive, his message is delivered only at the end of that turn.

Engineering: a division of engineers takes five turns to make a passage through a hedgerow, ditch, wall, etc. Ordinary infantry take ten turns. A division of engineers may plant a demolition charge (for a bridge, wall, etc.) in ten turns, and may then blow it when they wish. If a 0 or 1 is thrown on a nugget the charge fails to explode, and the work must be done again.

Artillery fire

Artillery batteries have the following maximum ranges, provided that they can see the target:

Horse artillery – 750 m
Field artillery – 1000 m
Heavy artillery – 1250 m

All artillery is assumed to be firing grape or canister when the enemy comes within 200 m, although morale tests may then force the guns to cease firing. When the enemy comes within 50 m during a turn, the gun's fire effect will not be calculated in the normal way, but as part of the combat rules (see below).

Every time a battery fires, read off the status lost (in tenths of a grade) by the target from the table below, and note the change on the target unit's record sheet. The scores shown in brackets are the effect when canister is used. For example, a 'B' grade field battery firing at 'C' grade cavalry would reduce it to 'C' minus 0.3 at long range, or 'C' grade minus 0.6 at canister range.

Loss of status to units under fire by one battery for one turn

FIRING UNIT		TARGET TYPE AND FORMATION		
Bty type and status		*Cavalry* *Inf. column/square* *Limbered arty*	*Inf. line* *Deployed arty*	*Skirmishers* *Prone inf.* *Troops in cover*
Field arty	'A'/'B'	0.3 (0.6)	0.2 (0.4)	0.1 (0.2)
	'C'/'D'	0.2 (0.4)	0.1 (0.2)	— (0.1)
	'E'/'F'	0.1 (0.2)	— (0.1)	— (0.1)
Horse arty	'A'/'B'	0.2 (0.4)	0.1 (0.2)	— (0.1)
	'C'/'D'	0.1 (0.2)	— (0.1)	— (0.1)
	'E'/'F'	— (0.1)	— (0.1)	— (—)
Heavy arty	'A'/'B'	0.5 (1.0)	0.3 (0.6)	0.2 (0.4)
	'C'/'D'	0.3 (0.6)	0.2 (0.4)	0.1 (0.2)
	'E'/'F'	0.2 (0.4)	0.1 (0.2)	0.1 (0.1)

Variations to artillery fire effects

Where a blank is shown in the above table, throw a nugget. The target will lose 0.1 from its status if the score is 7, 8, or 9.

When buildings are under fire from artillery, throw a nugget for each battery firing. If a 9 comes up, the building has started to burn, and must be evacuated at once by all its occupants. This represents the fact that each battery contains a number of howitzers capable of firing explosive shell.

Enfilades Infantry in line or deployed artillery count as infantry in column or limbered artillery if the firing unit can enfilade them, i.e., if it is firing at 90° to their line, so that its shot can rake their whole length.

Troops in cover or prone count as 'in line' if artillery can enfilade them properly. In reality, not even good cover was totally effective against this type of fire, and enfilading became a principle weapon in fortress warfare.

Hard cover In certain cases (e.g., troops behind very solid masonry) cover will be especially resistant to artillery, so that the effects of fire will be halved. Players must agree before the game which parts of the battlefield count as hard cover. Note that most buildings in the Napoleonic period count as soft cover, since they were usually quite flimsily built, of wood, wattle and daub and so on. Buildings also gave off nasty flying splinters under artillery fire, which increased the casualties among those inside or near them, thus partially counter-balancing their use as 'cover'.

Artillery which has been manhandled or traversed for half a turn may then fire for half a turn, but at half normal effectiveness.

Staff officers or individuals: whenever they are in danger, either from artillery fire or if they are entangled in a close combat, a nugget is thrown: they are killed for a score of 0; wounded for a score of 1 or 2.

Musketry

Musketry ranges are as follows (in metres):

Type	*Maximum range*	*Effective range*
Normal muskets	200	50
Rifles	400	100
Cavalry carbines	100	10

All divisions which are in view of the enemy within these ranges may fire, provided that they have not already moved on that turn. Note that only the front rank of models may fire: models further to the rear may not fire over the heads of others, because one line of models represents a triple line of soldiers, and fire could not be delivered by more than two or three deep. Units which receive an enemy attack during a turn may not fire in the normal way, their fire is taken as part of the general calculation for the close combat (see below).

One nugget is thrown for every division (or cavalry squadron) which fires. If a score equal to or higher than that shown in the table below is thrown, the target loses half a status grade; for example, four divisions of formed troops at status 'C' firing at formed troops in the open at long range roll scores of 0, 6, 8, and 9. This means that the enemy lose half a status grade, for the 9. If the same fire had been at short range, the enemy would have lost one and a half status grades, half a grade each for the 6, 8 and 9.

Dice rolls needed to inflict half a grade status loss on the enemy

FIRING UNIT		TARGET			
Formation	Status	Formed troops in the open		Skirmishers, gunners, or formed troops in cover	
		Long range	*Short*	*Long range*	*Short*
Formed	'A'/'B'	9	5	–	7
	'C'/'D'	9	6	–	8
	'E'/'F'	–	8	–	9
Skirmishing	'A'/'B'	8	4	9	6
	'C'/'D'	9	5	–	7
	'E'/'F'	–	8	–	9
Formed or skirmishing, any status except 'shaken' or 'disorganized'		–	8	–	9

Close combat

When opposing units come within 200 m of each other, both must test morale (see above). If an attacker survives this and continues to advance to within 50 m of an enemy unit, there will be a close combat. This does not normally mean that there is true hand-to-hand fighting, except between two cavalry units or in street fighting. It usually means that when they come within that range of each other the two sides will study the form very closely indeed, and one of them will be outfaced.

Skirmishers are automatically pushed back to rejoin their parent unit, without loss, if they are attacked by a formed body. Skirmishers beyond 400 m from their parent unit, however, will not be able to rejoin it unless they roll 9 on a rallying dice.

Exceptions to this are when the attacker is himself 'shaken', when he would not have been able to make an attack in the first place, or in woods, when the attacker will himself be automatically in skirmish order. When attacked by other skirmishers, as in woods, skirmishers do not retreat, but stand their ground. The combat then continues as a fire fight until one side or the other decides to retire. Notice that in thick woods both sides will also automatically be 'disorganized', so they will fire at lesser effectiveness.

Equally, skirmishers may advance to within 50 m of a formed unit, but must instantly rejoin their parent unit if that formed body makes the slightest offensive movement. Skirmishers may not press home an attack on anyone: their action is entirely by fire.

When formed units are in close combat with a formed enemy, apart from cavalry against cavalry (see below), both sides halt when their first ranks are within 50 m of each other. Combat is resolved as follows:

Both players agree which units are in each combat, and in what formations. At this point players may make attempts at quick rallies for any unit which is temporarily 'shaken' or 'disorganized', including any attacking units in column which have 'bunched' too close to their neighbours in the approach march. Failure to rally 'bunched' units will disqualify them from being counted in the combat, although they must share in its result; i.e., if they do not rally, their combat points will not be added to the total score of that side, but if that side loses the combat, they will also be deemed to have lost. If the defender claims any crisis initiative moves, they must also be settled now.

Units which are following too close behind either attacker or defender (i.e., units which have 'bunched' to less than 50 m behind front line units) must be counted in the result of combat, but will not be counted in the combat scores; they will simply be sucked along in the combat without being able to influence it.

Apart from these exceptions, all units which naturally fall into the same action, all the battalions attacking the same line, etc., are included in the one combat. Players must agree where the boundaries fall between each separate combat.

A small Austrian force, deployed in line, has been attacked by a mass of French infantry, in several lines of battalion columns. The Austrian cavalry on the left flank has charged the infantry and forced two battalions to form square, thus upsetting the attack. This has disrupted the cavalry's first echelon. On the French left a battalion in a column of double divisions is coming under fire from the deployed Austrian line. The Austrian battery is firing into the juicy target presented by the French square. The French skirmishers, meanwhile, are retiring in the face of the cavalry threat.

The central square has succumbed to the artillery fire and the cavalry has been reinforced. The Austrian cavalry's first line is still in disorder, mixed with fleeing French infantry. On the French left, their lead battalion has been checked by the Austrian infantry, but on their extreme right flank a fresh battalion column is still advancing around the square which remains.

Both sides then add up the total point values of all their units which are allowed to count in the combat, according to this table (any units still temporarily 'disorganized' or 'shaken' will count as 'E' grade):

Unit type	*Status*					
	'A'	'B'	'C'	'D'	'E'	'F'
Artillery battery	4	3	2	1	0	0
Battalion	5	4	3	2	1	0
Heavy cavalry regiment	6	5	4	3	2	1
Light cavalry regiment	5	4	3	2	1	0

Both sides now make the following additions or subtractions for each unit's score:

ADD ONE to combat score for:

All units making an attack or counter-attack.

Any one unit closely accompanied by the Division commander, who will be entangled in the fight.

All defending units on ground higher than the attacker.

All defending units behind normal cover.

The cavalry now meets the French second line, and finds that it comes under fire from a new square hastily formed by the French battalion to their right, as well as a column of double divisions to their front, and the original square to their left. A desultory firefight continues between the Austrian infantry and the shaken French battalion whose original attack failed. Unless they can commit more troops to the battle, the Austrians will now find that the tide of battle is turning against them, as their cavalry has been defeated, and they are heavily outnumbered in the infantry battle.

All lancer units fighting against infantry or artillery, for their longer weapons.
All armoured cavalry units.
All units larger than normal, e.g., five-division battalions, five-squadron cavalry regiments.
Add two to combat score for:
All defending units behind hard cover.
Deduct one from combat score for:
All units which have been in any close combat on the immediately preceding turn.
All infantry not in line to receive an infantry attack.
All cavalry units attacking, from any side, against infantry in column.
All cavalry units attacking frontally against infantry in line.
All units smaller than normal, e.g., three-division battalion, three-squadron cavalry regiment.
Deduct two from combat score for:
All cavalry units on foot.

DEDUCT THREE from combat score for:
All cavalry units attacking infantry in square.
Any unit in column of route, including units trying to storm across bridges.

After all this, both sides will have an overall total for their units in the combat. Both sides throw one nugget to find the variations in score due to chance:

Nugget score	*Result*
0	combat score unchanged
1	add 25% to combat score
2	add 50%
3–6	add 100%
7	add 150%
8	add 200%
9	add 400%

The side with the higher final total is the winner of that close combat. If the result is a draw, the defender wins.

The losing side must withdraw all units engaged in the close combat one full move at once (free movement), and all these units are automatically temporarily 'shaken'. In addition, a nugget is rolled for each of these units, to see what status they have lost. This result should, ideally, be kept concealed from the enemy.

For nugget score of 0 or 1	unit loses 2 status grades
2 or 3	unit loses 1.5 status grades
4 or 5	unit loses 1 status grade
6 or 7	unit loses 0.5 status grade
8 or 9	unit loses nothing

One grade is deducted from the nugget score if there is no unit of the same brigade in support behind that unit. Two points are deducted if there are no supporting units at all behind the unit.

The winning side advances 50 m at once (free movement) and stops on the enemy's former position, without suffering any loss. If the winner's force included any cavalry units, they must throw a nugget to test morale. If they lose the morale throw they recklessly pursue the enemy, and must throw for losses, as he does. They will remain confused with the enemy unit during his immediate retreating move after combat, and may disengage on the following turn.

In the case of cavalry combat against other cavalry, the same formula is used as before, except that the assessment of results is broken down into two phases, to represent the succession of clashes as first the front line is engaged, and then the second. In the first phase all regiments which are within range of the enemy commit their front line squadrons, and these become involved in a battle.

Regardless of type, status or tactical position, the side which has committed numerically more squadrons in the front line now automatically advances 50 m to the enemy starting position. The enemy must now test morale, and if he loses the test he also loses the combat as a whole. If he passes this test, on the other hand, he will be fighting at an advantage, having lured the opposing front line on to his own fresh reserves. In this case the combat is settled in the normal way, but with both sides deducting one point from their final combat scores for every squadron they had placed in their original front line.

To give an example of this process, the Royal Wallamaloo Yeomanry, light cavalry, status 'C', attack with three squadrons in their first line against the Cuirassiers de Cornichon, armoured heavy cavalry, status 'B', who have only one squadron in the front line. In the first phase the yeomanry automatically push back the enemy 50 m, and he throws for morale. He throws a 3, which is just enough to continue the fight into the second phase.

Both sides must now add up their combat scores in the normal way. The yeomanry have a basic score of 3, to which they add one for making the attack, but deduct three for their original front line strength, giving a total of 1. The cuirassiers, on the other hand, have a basic score of 5, to which they add one for their armour, but deduct one for their original front line strength so the total is still 5. These scores are now adjusted according to a chance dice: the yeomanry throw 7, which adds 150% to their score; $1 + 1.5$ giving a final total of 2.5. The cuirassiers throw 2 which only adds 50%; $5 + 2.5$ giving a final total of 7.5. This, however, still gives them a very convincing margin of victory, and the yeomanry

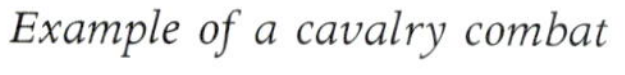
Example of a cavalry combat

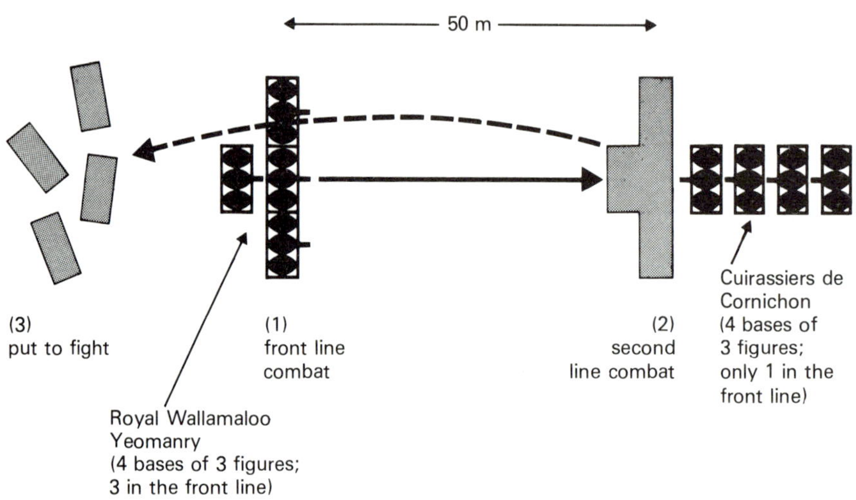

A contemporary print of the epic encounter between the Royal Wallamaloo Yoemanry and the Cuirassiers de Cornichon. Notice that both sides have trained their horses to march in step!

must withdraw, 'shaken'. The cuirassiers still have to test for morale again, to see whether they pursue recklessly. A lucky roll of 9 shows that they are totally in hand, so it remains only to see what the yeomanry have lost. The nugget they roll for losses shows 5, so their status is reduced from 'C' to 'D'.

4
The brigade game

The perceptive reader may by now have realized that there is a certain gap between the skirmish game and the Divisional game. The skirmish game deals entirely with details and individuals, while the Divisional game deals only with large masses, regiments and battalions. Nowhere do we really get a full sense of exactly what happens inside a battalion. It is to fill this gap that we now turn to the brigade game.

The Napoleonic brigade battle

Napoleonic brigades could vary enormously in composition. In Wellington's army they consisted of four or five battalions and a rifle company or two; while in the Prussian army they were almost indistinguishable from what other people called Divisions. In our games, therefore, we may use a handful of battalions, with perhaps a battery of artillery and a squadron or two of cavalry. This can be varied at will, to suit the particular needs of each specific game.

It was even less common for isolated brigades to meet the enemy than it was for isolated Divisions. Brigades would be much more likely to fight as part of a Divisional, corps, or army battle, although in some cases the connecting links might admittedly wear somewhat thin. Wellington's Light Division in the Peninsular War, for example, frequently spread its brigades over a very wide area, and demanded a high level of self-sufficiency from them. In the main, however, it was far more normal for brigades to fight with plenty of support nearby. If they were overwhelmed they could be replaced in the front line by fresh troops, and the battle could continue. Brigades would thus usually fight with a higher headquarters fairly close at hand.

The brigade commander's task was in some ways rather similar to the Division commander's, since both often operated with several lines of battalions which had to be fed in successively at the right moments and in the right places. Unlike the Division commander, however, the brigade commander would have to exercise a great deal of personal control and supervision of his men. He would have to command drill movements, and watch over the morale of his units. He would no longer be a rather remote figure, but would be well known to his troops. In order to design our brigade game properly, therefore, we must give the commander this added responsibility for what goes on at low level.

What you will need to play the brigade game

You will need a playing surface and scenery as described for the two previous games, and stationery.

Two model brigades

Brigades should be made up of 15 mm or 25 mm figures, exactly as for the Divisional game, and mounted on the same size of bases. The only difference will be that in this case one model soldier represents ten real men instead of thirty-three, and the ground scale will be 5 mm representing 2 m, instead of 5 mm representing 5 m.

These scales mean that each base of five model infantry figures represents a section, which is half a company, instead of a division of two companies, as in the Divisional game. One base of three cavalry figures represents a section, which is half a platoon or a quarter of a squadron, instead of a whole squadron. Each model cannon represents a section of two model guns, two limbers and their associated teams, instead of a whole battery. Each model caisson represents four real caissons or other battery vehicles.

Organization of wargame units for the brigade game

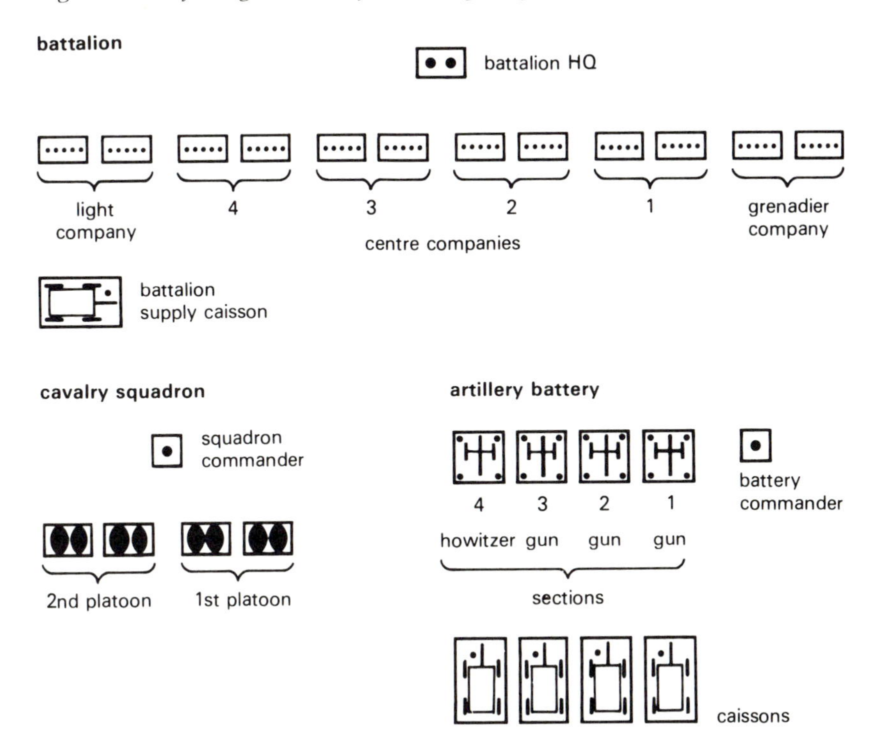

To put this another way, a battalion is now represented by about twelve bases instead of four, assuming that it has six companies; a squadron of cavalry is represented by four bases instead of one; and an artillery battery is represented by three or four cannon plus three or four caissons, instead of simply by one cannon on its own. Every battalion, squadron and battery must have a mounted figure to represent its commander; and the brigade as a whole should have a model staff group to represent its headquarters.

Playing the brigade game

A scenario is devised on very much the same lines as for the Divisional game, but bearing in mind that with the scales used for the brigade game, each square inch of the model battlefield will represent considerably less real ground. Thus the units will have much less elbow room on the table top than in the Divisional game, unless a bigger playing area, such as a floor, is used.

Orders

One of the most distinctive features of the brigade game is the importance attached to the transmission of orders. Very little may be done by units unless they have very specific orders: for example, 'advance over that wall and capture the farm'; 'form square'; 'open fire', and so on. It is therefore essential that all the stages in the transmission of orders are followed by the player. The skill in this game consists in anticipating what orders will be required, perhaps two or three turns ahead, so they can be transmitted in good time. The player who masters this skill better than his opponent will usually be the winner.

At the start of each turn the player, who is represented on the table by a specific figure, may write up to two orders. He then works out how many turns it will take for these to be received by units. For example, it may take a courier one turn to take the message from the player's figure to the recipient battalion commander; and the battalion commander takes a turn to pass the message down to his individual sections of soldiers, according to the rules for movement. In this case the order would take two turns to transmit, plus half a turn to issue, making a total of two and a half turns. When the player writes the order he must also make a note of the time at which it will start to be obeyed.

Only very minor actions are allowed outside the written orders, mostly common sense interpretations of them. If there is any dispute, the unit concerned will be allowed to act against orders only for a nugget roll of 7, 8 or 9.

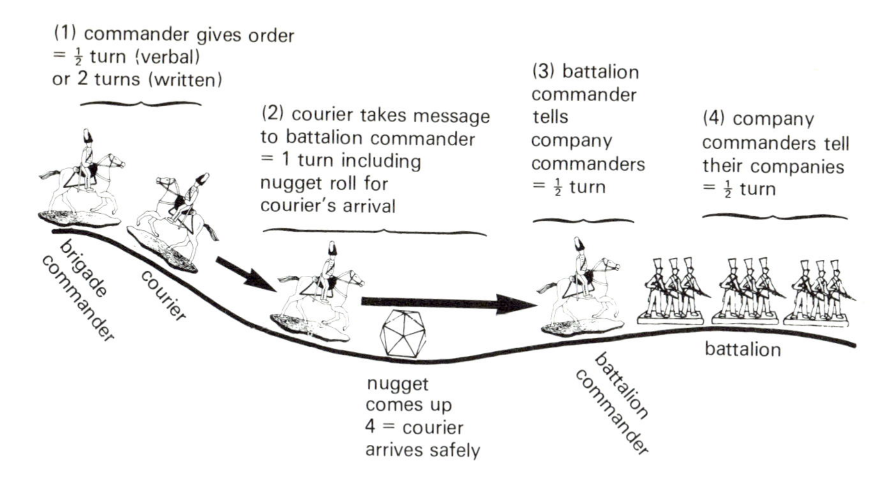

Example of transmission of orders in the brigade game

Scales

We have already stated that the figure scale is one model to ten men, and that the ground scale is 5 mm to 2 m. The rules for the vertical scale use similar calculations to the Divisional game, while the timescale in the brigade game is: one turn or bound equals half a minute. This means that three times as much action may take place per bound as in the skirmish game; but only a quarter as much as in the Divisional game.

Status and losses

As in the Divisional game, it is very important to master the system used for status, losses, and morale. Each unit is given a consolidated status grade; but in this case there will be a separate status for each individual section, whether of horse, foot or gun, instead of for each whole battalion, battery or squadron. Thus each section will start the game at a status of 'A' for regular, élite, or flank companies; 'B' for regular centre companies, or reserve flankers; 'C' reserve centre companies; or 'D' for irregulars, etc. Each status grade lost during the game represents the loss of about five men, to all causes; every three grades lost by an artillery section will see one cannon written off.

Sections which fall to 'E' status are automatically and irretrievably 'shaken', while those which fall to 'G' status are removed from the table.

Morale

Each section may sometimes be temporarily 'shaken' or 'disorganized', before it falls to 'E' status. Because morale is assessed separately for each section within the battalion, however, this means that some parts of a battalion may be 'disorganized' or 'shaken', while the rest of the battalion is perfectly unaffected. It is this, along with the cumbersome

Although the terrain at Austerlitz does not look like this, the picture gives an excellent idea of the extended lines used in Napoleonic battles. In the centre, just behind Napoleon's head, we see one battalion deployed in line, with its mounted officer and N.C.O.s in the fourth rank. Further afield we see a regiment of four cavalry squadrons attacking to the left, and an artillery battery of five guns deployed on the right, with caissons parked behind. Note also the characteristic scattering of stragglers around the battlefield – the men who failed to keep their alignments.

procedure for passing orders, which gives the brigade game its special character. It is rare for any single battalion ever to be completely unaffected, so commanders must spend a great deal of time and energy trying to keep their lines steady, exactly as in real life.

A disorganized section is perfectly under control, and may fire accurately, and so on, but it is not in a good position for close combat. Whenever a section passes through rough ground or obstacles, or whenever it skirmishes, runs or changes formation, it becomes 'disorganized' and loses its accurate alignment. This is also true (unlike in the Divisional game) whenever a section fires its weapons. It always requires one whole turn, stationary, to retrieve a section's alignment; what the drill sergeant calls 'taking your dressing'. Skirmishers and gunners are always counted as 'disorganized'.

Sections become temporarily 'shaken' when they fail a morale test. This means that they automatically stop whatever they are doing, and open fire wildly at any enemy in sight, however far away he may be. To stop a shaken section from firing requires a rallying dice, and even that will not in itself allow the unit to regain its 'dressing'. Even if there is no enemy in sight a 'shaken' section will not be allowed to make any advance, although it may still be ordered to perform other drill movements. Once again, the section will remain 'shaken' until it has been rallied by a favourable rallying dice.

Morale tests are taken in the following circumstances:

The section sees the enemy for the first time.

The section comes within 200 m of the enemy or the enemy approaches within 200 m of the section.

Whenever there is a change of orders while the enemy is within 200 m; this is for formed infantry only and does not apply to cavalry, artillery, or skirmishing infantry. It includes the order to open fire, since there were few actions which so disturbed troops on the Napoleonic battlefield as having to use close order fire.

A friendly section tries to pass through the section.

The battalion commander is killed or wounded within voice range, 30 m.

The adjacent section is 'shaken' from the previous turn, and has not been rallied on this turn. This allows unease to spread down a line of troops, unless it is checked early.

In combat (see combat rules, below).

The morale test itself consists of rolling one nugget. The section fails the test and becomes shaken if the following scores, or less, appear:

Section status	*Nugget score*
'A'	1
'B'	2
'C'	3
'D'	4

Some possible combinations of morale within a battalion

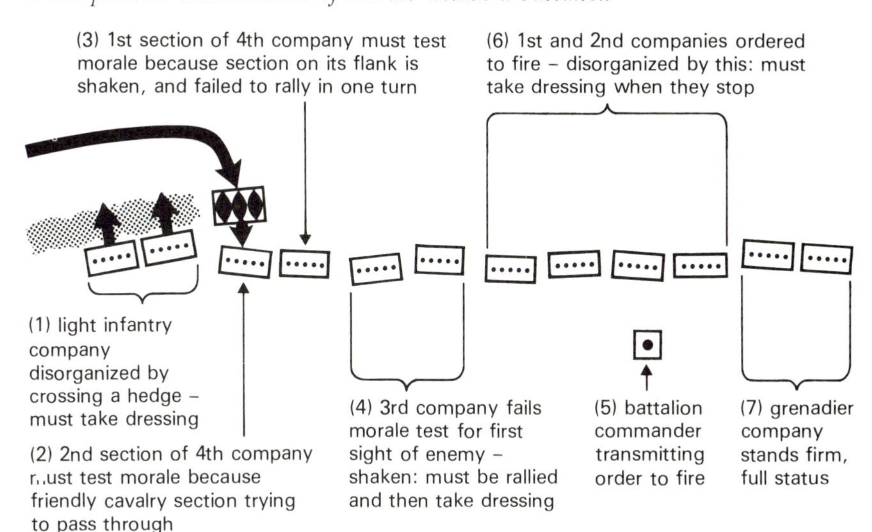

Rallying a shaken section is achieved by rolling a 9 on a nugget. One attempt at rallying may be made each turn.

Sequence of actions in each turn

(i) All players make written notes of all orders which they are supposed to issue on that turn (up to two per turn), and a note of the turn on which the order will start to be obeyed.
(ii) Both sides simultaneously move. If this includes any morale tests, they are taken at this stage.
(iii) The result of any firing is calculated.
(iv) The result of all close combats is assessed; and any attempts to rally units.
(v) Both sides agree that the turn is over and the next may begin.

Command

Commanders may issue up to two verbal orders in a turn, i.e., either two couriers may be briefed and sent out from the player's position, or two direct commands may be issued to a battalion, battery, or squadron commander, provided that his figure is within voice range, 30 m.

A written order takes two turns to write and dispatch. If there is anything complicated, e.g., which involves a sketch map or a complex sequence of actions, it must be written down, unless the recipient is within voice range. After they have received verbal or written messages, couriers move at 120 m per turn, but must throw a nugget: if 0 or 1 is thrown, they move at quarter speed.

When orders are received by subordinates, e.g., battalion commanders receiving a courier from the brigade commander, they may be passed on down the hierarchy at the rate of one level every half turn. Thus it will take a battalion commander half a turn to brief his company commanders provided they are within vocal range, and they will require a further half turn to brief their companies. Signals may be used for relays beyond vocal range, provided that they have been arranged before the game. They take as long as verbal orders to relay.

Infantry

Once an infantry section has received orders from its company commander it may start to act. During each turn it may:

Walk 50 m on good going, in good order; or 30 m on bad going, in 'disorder'; or 10 m backwards or sideways, in 'disorder'.

Charge or run 70 m, in 'disorder'.

Cross obstacles, low walls, hedges, fences, etc., in 'disorder'.

Lie down, stand up, fix or unfix bayonets, etc., all in 'disorder'.

Load and fire any weapon but a rifle; a rifle takes two turns to load and fire. Loading and firing must be done stationary, in 'disorder'.

The basic formation for the brigade game, of a battalion in square. Note that in the square formation, the battery vehicles take refuge inside the hollow square, along with the battalion staff, flags, drums and so on.

Once troops are ordered to deliver fire in close formation, they must take a morale test. If they become 'shaken' as a result, they must be rallied before they may cease fire (see morale rules, above).

Change formation Each section moves at normal speed, once specific orders have been received to change formation, in 'disorder' for the duration of the change.

Take dressing Any 'disordered' unit which is not 'shaken' may be returned to full alignment in one turn, provided it is stationary and not doing anything else. If it is 'shaken' it must be rallied before it may take its dressing.

Cavalry

Once the cavalry has received orders from its squadron commander, it may act as follows in each turn (distances in metres):

	Walk	*Trot*	*Gallop*
Good going	70 (in good order)	90 (in good order)	Heavy cavalry – 100 (in disorder) Light Cavalry – 120 (in disorder)
Rough going	50 (in good order)	70 (in disorder)	80 (in disorder)

Note that a standing section must always walk one turn before it may trot, and a walking section must trot one turn before it may gallop. Equally, a galloping section must trot one turn before walking, and a trotting section must walk one turn before stopping. If stopped by a morale test, trotting or galloping cavalry do not stop at once, but turn away and slow down later.

Taking jumps, one turn in 'disorder' plus roll one nugget. Lose one status grade for 4, 3, 2 or 1. Lose two grades for 0.

Mounting or dismounting, takes one turn, in 'disorder'.

Changing formation, as for infantry.

Artillery

Artillery moves as follows (in metres) each turn, provided orders have been received:

	Horse artillery	*Field artillery*
Good going, limbered	70	50
Bad going, limbered	50	30
Manhandled	20	10

Limbering, unlimbering, or opening caissons take one turn each.

Preparing to fire, after unlimbering, takes one further turn for medium guns, two turns for both heavy and horse guns (the extra horses get in the way).

Loading and firing one shot takes one turn, except for very heavy guns, when it takes two turns.

Traversing artillery through more than 45° takes half a turn.

Note that artillery may not cross obstacles, and that if they are attacked, gunners will always count as being in 'disorder'. This 'disorder' does not reduce their ability to move or fire, unless they are also 'shaken'.

Firing

Ranges are as follows:

Weapon	*Maximum range*	*Effective range*
Horse arty	750	200
Field arty	1000	200
Heavy arty	1250	200
Musket	200	50
Rifle	400	100
Carbine	100	24

Find the range to the target, the status and morale of the firing unit, and the profile of the target. A nugget is then rolled for each section firing, and the score is adjusted for certain variables:

All artillery always adds two to its dice score.

If the firing unit has not fired before in this game, add two.

If the firing unit is mounted, deduct two.

Artillery firing in enfilade throws two dice per section, not one dice.

Infantry weapons may not fire in rain, but artillery may.

Now compare the final score with the table below. If the score equals or exceeds the score shown, there is one hit on the target, i.e., one section loses one status grade.

FIRING UNIT'S STATUS	TARGET'S PROFILE		
	Inf. column/square Cavalry Limbered arty	Inf. single line Unlimbered arty	Skirmishers Inf. in cover
Long range			
'A'/'B'	8	9	10
'C'/'D'	9	10	11
'Shaken' or status 'E'	10	11	12
Effective range			
'A'/'B'	6	7	8
'C'/'D'	7	8	9
'Shaken' or status 'E'	8	9	10

Close combat

As the distance between two opposed forces decreases, there will be morale tests when any fresh units first see the enemy. Then, when the distance has closed to 200 m, both sides take a morale test whether or not they have been in action before. If the attacker can still keep enough of his force in hand, or rally them, he may continue to advance. When he comes within 50 m of the enemy, the close combat sequence is used.

Both sides agree on how many sections are involved in each combat. No combat should normally have more than one battalion, battery or squadron per side.

Now find how many sections in each combat are enthusiastic to continue the fight. Throw a nugget for each section in the combat, and make the relevant adjustments for the tactical circumstances:

Cavalry which can find the flank of an infantry or artillery force add 3 to their score.

Defending infantry and artillery which are behind cover or uphill from an attack add 1 to their score.

Attacking troops coming downhill or taking the enemy in a flank add 1.

Any unit within voice range of the brigade commander adds 1.

After finding the total for each section, compare it with the scores shown below. Any score which equals or exceeds the score shown represents a section which is still interested in the fight:

	Status				
Unit	'A'	'B'	'C'	'D'	'E'
Fully formed/in hand	5	6	7	8	9
'Disorganized'/'shaken'	8	8	9	9	10

Now add up the total number of interested sections on each side, and make a note of which they are. The side with the higher number is the winner of the combat.

If the attacker wins, he occupies the enemy position, but is 'disorganized' until he can take his dressing. All his sections which failed the enthusiasm test in combat must now test for morale in the normal way. Any section which fails this test becomes shaken and loses one status grade. The defeated defender must also retire one move, 'disorganized' and must test morale for all his sections. Again, all sections which fail this test become 'shaken', and lose one status grade.

If the defender wins, he holds his ground, but is 'disorganized' until he takes his dressing. Those sections which had failed the enthusiasm test must now test for morale in the normal way: any which fail become 'shaken', and lose one status grade. As for the defeated attacker, he must stop 50 m short, 'disorganized', and must test morale for all his sections. Any which fail become 'shaken' and lose one status grade.

If there are any cavalry sections among the winning side, they must all automatically test morale. If they fail, they become 'shaken', lose one status point and pursue the enemy recklessly.

5 The army level game

We cannot discuss games with model soldiers without also mentioning the highest level of them all, that of the army or corps. At this level we portray an entire Napoleonic battle in one game, and are no longer forced to play only a part of a battle in which the result will ultimately turn on external circumstances and the decisions made by some higher headquarters. When we play the army level game we are at last taking the role of the higher headquarters itself; we can be Wellington at Waterloo, the Archduke Charles at Aspern–Essling or Napoleon at Austerlitz.

There are two ways of setting up a game at army level. Either you can spend two or three days playing a series of Divisional games simultaneously in a large hall or even, in theory, a huge multiplicity of brigade games spread over a month or two, or you can accept simpler rules and play a single, unified game within a few hours. This second alternative means that in many details the rules will inevitably be much less realistic than those for the games we have discussed so far; but in the central fact of portraying high command they will actually be more realistic. If the player has to take the same sort of decisions as Wellington or Napoleon, within about the same space of time, it will obviously make a better simulation than if he has to make a lot more decisions, at a lower level, spread over a longer time.

The Napoleonic army battle

The majority of Napoleonic battles involved either an army corps on each side, or a whole army. This is really just another way of reiterating my earlier statements that very few battles consisted of isolated Divisions on their own. The somewhat static and formal Peninsular battles were fought with several Divisions on each side, while in central Europe there tended to be several army corps, manoeuvring over a frontage of many miles. The action would often start when an advanced Division bumped into the enemy. The corps commander might then concentrate all his forces upon that point, and a corps action would be fought. If the rest of the army was relatively widely dispersed, then the affair might not develop any further than that, since reinforcing corps would be unable to arrive before a decision had been reached. This was the case, for example, at the battles of Golymin (1806), Kulm (1813) and Vauchamps (1814).

If the main armies were concentrated, on the other hand, the corps battle would soon be reinforced and taken over by army headquarters. This would be the signal for a general army action to develop, which might take between eight hours and three days to complete. No two were the same, of course, but it is worth mentioning some of the general characteristics which have been considered typical in the past.

At the start of the battle the first units to arrive would usually engage the enemy frontally and try to pin him. Reinforcements would be fed into the line as they arrived, in an attempt to stabilize the position and maintain pressure on the enemy. The army commander might arrive at this stage, and make a reconnaissance of the ground and the possibilities. A great deal of what happened subsequently would depend upon his appreciation of what was going on.

The skill of the commander would be applied to drawing the enemy's reserves into the action while husbanding his own, for once a reserve unit was committed, it was very difficult to extricate it and use it on another line of action. It was also important to identify the enemy's line of communication, and try to threaten that. Army corps which had not yet arrived on the battlefield might therefore be sent hectic orders to swing wide, and try to arrive on the enemy's flank or rear. Once again, this would keep up the pressure on the enemy's reserves, and it was often this sort of concentration on the battlefield which held the whole key to success.

The stages in the development of an army's battle (according to H. Camon, 1910)

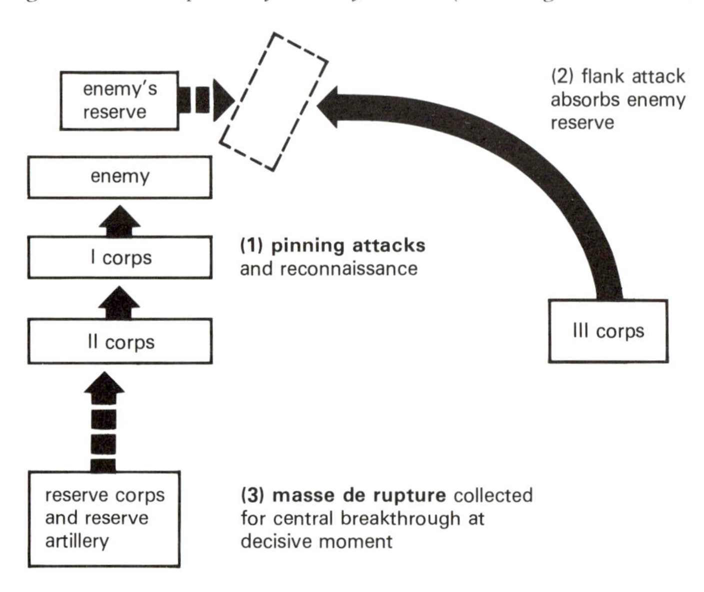

As the battle developed, the successful commander might gradually collect a *masse de rupture* in reserve. He would eventually throw this forward in a huge frontal attack, supported by massive artillery fire, at the moment when he judged the enemy to be sufficiently worn down. If everything went well this mass would break through, and its cavalry would fan out in pursuit of the beaten enemy.

Things could go wrong at any stage in this process, however. The enemy might luckily defeat some of the early attacks, and thus be able to keep up more pressure than *vice versa* (e.g., at Eylau, 1807, when Augereau's corps was defeated early on). Again, a corps detailed to swing round the enemy's flank or rear might well arrive late or in the wrong place (e.g., Ney's corps at Bautzen, 1813). The final central attack might totally fail to break through (e.g., Waterloo, 1815); or might achieve only partial success (e.g., Wagram, 1809). Most likely of all, perhaps, the victorious army might be too exhausted to exploit its success in a pursuit (e.g., Borodino, 1812). In the end result, the plan would probably have to be changed several times, and would not look nearly as neat and tidy in practice as it had on paper.

What you will need to play the army level game

If you are fighting on a table top, you will need terrain and scenery as already described, but on a much smaller scale. 15 mm might just be acceptable for minor actions; but 5 mm is really the scale to aim for. If you are lucky enough to have access to a large hall, on the other hand, you can use 15 mm and 25 mm figures quite happily; but you will need a huge number of figures. You will once again need stationery, as for the previous games.

Two model armies

These should be divided into army corps and Divisions, and then into brigades and battalions. For table top use each battalion should be represented by a single small base about 10 mm long, with only a few figures on it. The 5 mm blocks which were produced a few years ago are ideal for this; but something quite effective can also be made up from suitably painted matchsticks.

Artillery batteries are represented by a single model gun; and cavalry regiments are represented by a single block of cavalry figures, also about 10 mm long. The wargamer can also make up staff groups for Divisional commanders and above; wagon units, pontoon bridges, and other impedimenta. We will be operating at such a high level that all these things may at last come into the game.

If you are using 25 mm armies in a large hall, the organization is as for the Divisional game.

Markers for simultaneous movement

Each player should have a stock of cardboard markers in his army's colour, to indicate his intentions for each unit on each turn. There should be a large number of battalion (or battery, or cavalry regiment) markers, plus a set of larger ones for each brigade. They should indicate 'rapid movement', including the intention to charge home if possible (a straight arrow); 'cautious movement', avoiding combat if possible (a zigzag arrow); the desire to 'stand firm', firing if appropriate (a blank); or 'digging in' (a spade symbol).

Playing the game

A scenario must be devised. It is best to play games of this scale as part of map campaigns (see chapter seven); but failing that, the game should be started with two Divisions (or even corps, if it is a big game) already in combat. We assume that this is the moment when the army commander reaches the scene and makes his appreciation. His remaining units should be spread out in a fan behind the point of contact: some already laid out on the playing surface, and some off the table, each a stated number of hours' march away.

A major problem with this game is visibility, for these operations were so huge that no one man would ever have been able to overlook the whole battle as our wargamers are now doing. The use of an umpire may alleviate this a little, and conceal certain units until they would really be visible to the enemy. In general, however, we will have to accept an unrealistically high level of visibility in this game.

Orders

At the start of the game the army commander writes general orders for each of his corps commanders. If there are players for the corps commanders, they must also write orders for each of their Divisions. All action must respect these general orders unless they are overruled by *force majeure* or a new order.

Scales

As with the Divisional game, there are four scales.

The ground scale is 1 mm to 10 m, if 5 mm models are used; or 1 mm to 1 m if 25 mm or 15 mm models are used.

The figure scale is, one 5 mm block or base of figures (about 10 mm long) represents one battalion, battery, or cavalry regiment; or, if a 15/25 mm scale is used, the figure scale is as in the Divisional game.

The vertical scale is the same as for the Divisional game.

The time scale introduces the great innovation of this game: each turn represents a quarter of an hour's fighting. This means that we must totally abandon any attempt to portray low level tactical actions, such as the precise moment at which a battalion forms square, when it fires and so on, and we have to accept a highly generalized result for all combats.

A quarter of an hour was roughly the time it took for infantry to advance from covered positions into a close combat. It was also, again very roughly, the sort of time during which units might expect to be in the very thick of the fighting.

Status, losses, and morale

Each unit starts the game with a basic status between 'A' and 'D', exactly as in the Divisional game. As it suffers losses its status may fall through 'E' (automatically 'shaken', may not attack) to 'G' (totally spent, removed from the game). The exact state of losses must be recorded by each player, or by an umpire.

Units become temporarily 'shaken' whenever they become 'bunched', or lose a combat (see combat rules, below), or when they fail a morale test. This fact is then indicated on their record sheet. Morale tests are taken whenever:

Any unit sets eyes on the enemy for the first time in the game.

A friendly unit attempts to pass through the unit.

The enemy can claim a rear attack on the unit.

Cavalry win a combat, and may try to pursue wildly (see combat rules).

The morale test itself consists of rolling one nugget. The unit becomes 'shaken' for the following score or less:

For status 'A'	–	nugget score 1
'B'		2
'C'		3
'D'		4

Rallying: whenever a unit becomes temporarily 'shaken', a nugget is rolled for it on the next turn, and every subsequent turn. The unit will rally if the nugget shows 8 or 9.

Sequence of actions in each turn

(*i*) Players simultaneously place a movement marker face down by every unit, to indicate what move is intended on that turn. In the front line it may be necessary to use separate markers for every single battalion, but if this can be avoided it is far more convenient to use only one marker for each brigade. This will also – realistically – encourage players to think of their brigades as relatively indivisible formations.

(*ii*) Players now turn their markers face up and move their units as planned. Because the markers are there for all to see, there can be no disputes over simultaneous movement. Whenever two opposing forces come within 200 m of each other they must stop at once. This constitutes a close combat for the rest of that turn.

(*iii*) All firing is assessed.

(*iv*) Morale tests and close combats are assessed.

(*v*) Attempts are made to rally all temporarily shaken units.

(*vi*) Players take receipt of any written orders, and agree to move on to a new turn.

Movement

Units move the following distances (in metres) each turn, until they are stopped by an obstacle or the presence of an enemy within 200 m, when a close combat starts.

Type of unit	*Open ground*	*Woods*	*Deductions for minor obstacles*
Infantry	700	300	– 200
Manhandling arty	150	50	– 100
Field arty or wagons	500	–	– 450
Horse arty	1200	–	– 1100
Heavy cavalry	1200	300	– 500
Light cavalry and staff	1500	300	– 500

Couriers move at staff rate, but do not arrive if 0 is thrown on one nugget.

Roads add 150 m movement to all units using them.

Rivers must be graded as 'minor obstacles', 'major obstacles' (which take one turn to ford, except for guns and vehicles), or big rivers, which can be crossed only by couriers, who are drowned if a new nugget roll shows 0.

Limbering/unlimbering artillery takes one third of a turn.

Mounting/dismounting cavalry takes one third of a turn.

Changing formation takes no additional time. Units are always assumed to fight in the most appropriate tactical formation.

Engineering Villages may be strengthened in two turns by a battalion, or well fortified in six turns. In the open, fieldworks for a battalion are constructed in eight turns.

Engineers set a demolition charge in two turns, and may blow it when they wish, provided a nugget roll comes up anything but 0 or 1.

A pontoon bridge is set over a minor obstacle in two turns; over a major obstacle in twice the number of turns shown on one nugget; and over a big river in four times the number of turns shown on one nugget. Dismantling pontoon bridges takes half their building time. Note that a nugget must be thrown for all pontoon bridges once every eight hours. If it shows 0, the bridge is broken by natural causes.

Fire

Fire may be applied whenever a stationary unit can see a target in range, but is not engaged in close combat on that turn: i.e., artillery fire is normally counted against targets between 200 m and extreme range, but not at ranges shorter than 200 m, since that fire is already counted into the close combat calculations. In some cases infantry may also have targets presented at less than 200 m against which no close combat is possible, for example, over a river, or in woods. Normally, however, only rifle-armed infantry will get an opportunity to use fire separately from the close combat.

RANGES (in metres)

	Field bty	*Horse bty*	*Heavy bty*	*Rifle*	*Musket*
Maximum range:	1000	750	1250	400	200

Effect of fire For each battalion, battery or cavalry regiment giving fire, read off the result on the target from the table below. This will represent the status lost by the target unit after one turn's fire:

FIRING UNIT'S STATUS	TARGET				
	Field bty	*Horse bty*	*Heavy bty*	*Rifle*	*Musket*
Fire against troops in the open					
'A'/'B'	1.0	0.8	1.2	0.5	0.2
'C'/'D'	0.7	0.5	0.9	0.3	0.1
'Shaken', or 'E'	0.5	0.3	0.7	0.2	—
Fire against troops in cover					
'A'/'B'	0.5	0.4	0.6	0.3	0.1
'C'/'D'	0.4	0.3	0.5	0.2	—
'Shaken' or 'E'	0.3	0.2	0.4	0.1	—

Artillery fire against buildings: throw a nugget. The buildings will catch fire and must be evacuated for a 7, 8 or 9.

Enfilades Double the effect of artillery on targets hit in enfilade.

Skirmishers The action of a battalion's skirmish screen is not assessed separately from the action of the battalion itself. In woods, however, and whenever else units are told to skirmish *en masse*, the entire unit is assumed to be split up into small skirmishing groups. It must then fight entirely by its fire.

Close combat

Whenever two opposing forces come within 200 m of each other, they must both immediately stop their movement and become fixed. If there are intervals of less than 50 m between each unit in the line they will also automatically become 'bunched', and fight as if 'shaken'.

If the combat takes place in woods, both sides must automatically be in skirmish order, and fight by fire alone. In most other terrain, however, any skirmishing unit fixed by the attack of a formed unit will be at a great disadvantage.

Find the total combat score of each side by adding the points of every unit in the same combat. There may sometimes be as many as a brigade in each.

For a unit at status		count
	'A'	5 points
	'B'	4
	'C'	3
	'D'	2
	Skirmishing, shaken, or 'E'	1

Next make additions and subtractions, as follows:
ADD ONE to combat score for:
Unit making an attack.
Defending unit in cover.
Unit on higher ground than the enemy.
Any one unit accompanied by the Division commander.
Lancers against infantry or artillery.
Heavy cavalry.
ADD TWO to combat score for:
Defending unit in hard cover.
Cavalry unit if accompanied by infantry, against infantry.
DEDUCT ONE from combat score for:
Unit which fought a close combat in the previous turn.
Unit caught in the act of digging in.
Unit without rear support from the same brigade.
DEDUCT TWO from combat score for:
Cavalry on foot.
Cavalry attacking infantry.
Any unit caught on the march or in a defile (e.g., storming a bridge).
Unit without any rear support.

Both sides see what variation to their total score must be made for the action of chance:

Nugget score	*Result*
0	combat score unchanged
1	add 25%
2	add 50%
3–6	add 100%
7	add 150%
8	add 200%
9	add 400%

The side with the higher final total score is the winner of the close combat. The defender gets the better of any draws.

The losing side must withdraw all units from the combat one full move, during the next turn's movement. All these units are automatically temporarily 'shaken', and lose one status grade. The winner suffers no loss, except that cavalry must test morale. If they fail the test, they pursue recklessly and become 'shaken', losing one status grade.

6
The generalship game

The four games we have looked at so far use model soldiers to play out battles and skirmishes. This is fine as far as it goes, but it suffers from at least three important weaknesses.

We cannot always achieve as much tactical and command realism as we would like, because we always have to accommodate the model soldiers into the game. They sometimes distract us from achieving higher levels of realism.

The players can see too much of what is going on. Real Napoleonic generals would not have enjoyed such an open view of their battles, and would certainly not have been able to intervene with low level details, such as battalion formations, as often as we do with our games using models.

All these games are purely tactical, but high Napoleonic generalship was really about many other things as well as tactics; e.g., strategy, logistics, intelligence gathering, issuing and supervising orders, building morale and so on.

All this poses a problem: how can we create a game which gives a fuller representation of Napoleonic generalship? To do this we must step back from the table top and the model soldiers, and return to first principles.

Napoleonic generalship

The Napoleonic army commander on campaign was faced with the basic problem of laying on a favourable battle, or set of battles. The actual tactics of the battle itself were of very secondary importance beside this major strategic aim.

In order to lay on a favourable battle, the various army corps had to be set in motion along the right roads, and with the right timings. They also had to unroll a line of communication behind them as they advanced, consisting of a string of posts and depots, with one day's march between each. These depots would then have to be injected with rations, clothing, ammunition, reinforcements, and a host of other articles which were necessary to keep the army at high efficiency. No army could live off the land totally; and even the most veteran troops had to keep some sort of contact with the umbilical cord of their line of communication.

The depots on the line of communication would need to be supervised and kept in good working order, e.g., by personal visits; and care would have to be taken not to overload their capacity. In peacetime this could all be relatively plain sailing; but in wartime there were enormous difficulties, since it was impossible to plan anything very far in advance. The army had to manoeuvre according to what the enemy was up to, so much of the line of communication had to be improvised at very short notice indeed.

In order to decide where the enemy was and what he was doing, the commander-in-chief had to set up his own intelligence service for each campaign; there was no permanent one as we understand it today. He would also be in correspondence with his government and his own family at home, and with any allied governments involved in the campaign. He would have to encourage the troops by inspections, bulletins, and speeches; and he would have to organize his own personal affairs so that he was always in the right place, with enough food and sleep to keep going. Finally, if he did succeed in bringing the enemy to battle, he would have to make sure of a tactical victory.

All this obviously imposed a terrific strain upon a commander, and to do everything properly he really needed to be everywhere at once. As a generalization, we can therefore say that the good commander needed two main qualities. He needed the judgement and military knowledge to use his time to the best advantage, and to put high priorities on those things which were really important. And he needed the energy and personal force of character to get things done, once he had decided what things needed to be done.

If we are to make a wargame of all this, we must obviously find some way of representing both of these qualities in the system of play. In fact there are two main methods for doing this: one is the somewhat formal generalship game discussed in this chapter; the other is the 'free' and informal map kriegspiel described in the next chapter.

In the generalship game each player represents a commander-in-chief, and each turn represents twenty-four hours. During the turn each player must indicate on a game board precisely what his commander-in-chief will be doing during that day: resting, travelling, writing orders, haranguing his troops, and so on. The time is therefore broken down into its component parts in rather the same way as for the sequence of loading and firing, or obstacle crossing in the skirmish game. The successful player will be the one who has been able to allocate his time most profitably, and who has achieved a better mastery over the types of action open to him.

In order to make room for this rather detailed portrayal of what goes on in the higher command, we are forced to cut down still further on the lower level details. In the generalship game we therefore portray the theatre of war and the army only in very schematic form. Players will no longer be able to manipulate individual battalions, as they did in the army level game, but will have to consider a whole army corps as the playing unit. This, however, is a very small price to pay for the much enlarged realism of the generalship itself.

What you will need to play the generalship game

An umpire (or preferably two) is vital for this game. There should also be as many players as there are commanders-in-chief in the campaign. In the Waterloo campaign three would be needed: Napoleon, Wellington and Blucher. The umpires should have a desk for their map, and the players should also have one each, out of sight of each other but accessible to the umpires. Players might sit back to back in the same room, or they might be in separate rooms.

Each team, and the umpires, should have a copy of a specially prepared schematic map of the theatre of war. This may be based upon a real historical map (as published in many military histories), or it may be entirely imaginary. Whichever it is, it should show all the main roads in the theatre, plus one town (neither more nor less) for every day's march along each road. Each day's march may be between 5 and 9 leagues (20–36 km; $12\frac{1}{2}$–$20\frac{1}{2}$ miles), so every town shown on the map must be somewhere within this distance from the next ones. In other words a force marching out of any town by any road will always be able

Example of a map and counters

Unit counters:

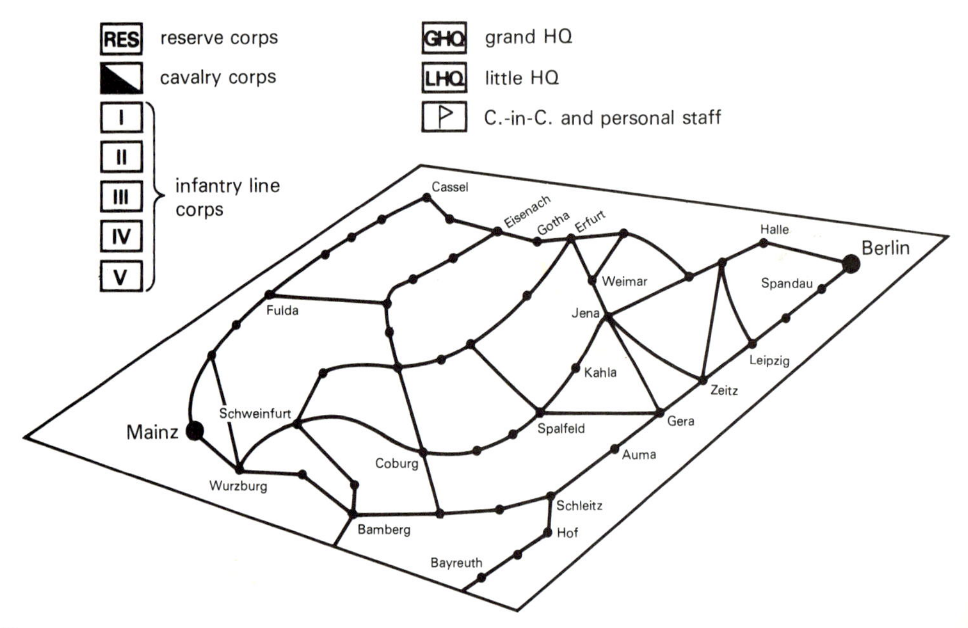

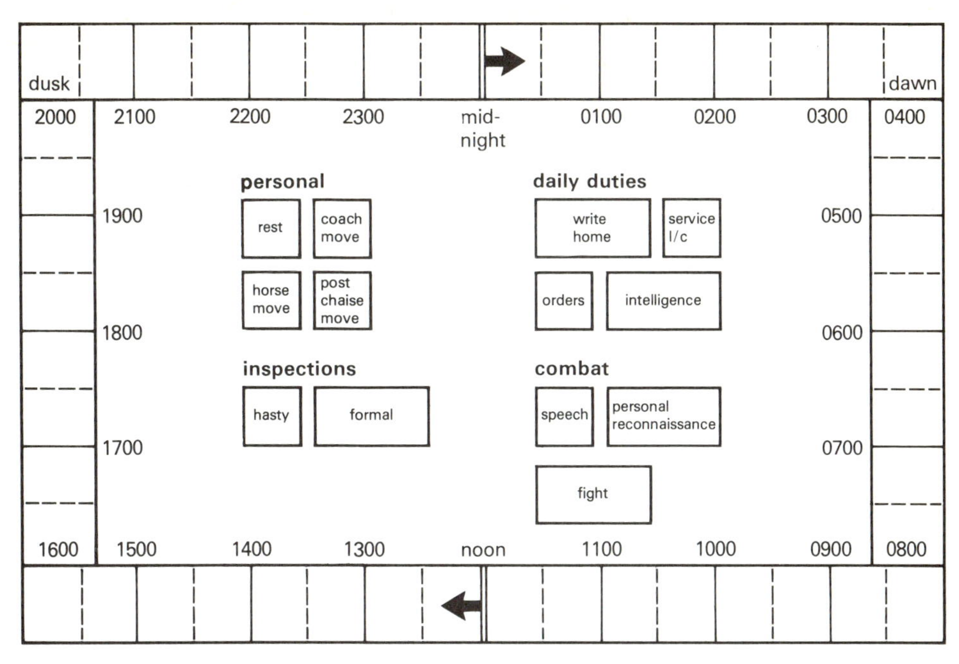

Example of a game board and action counters; there should be as many copies of each counter as will be needed in any move.

to finish its day's march in another town. The towns may be of any size, but for game purposes they are classified as either fortresses or open supply points. Each army will also have one town, usually a fortress, nominated as its base. The maps need not be to any particular scale, as all they have to do is show roads, towns and fortresses.

Each map should have a set of cardboard counters representing the subordinate formations of each commander. Every corps, infantry, cavalry or reserve, should have a separate counter; and there should also be one counter each for the commander and his personal staff, the little headquarters, and the grand headquarters. These units would in reality consist of approximately ten, five hundred, and five thousand people respectively, if we include the attached escort troops. The counters for every unit are moved on the map during play to indicate the positions of units. Any number may be stacked on any town.

Each team, but not the umpires, must have a specially prepared game board. This consists of a sheet of cardboard, usually 60 cm^2 (2 ft^2), around the perimeter of which are marked forty-eight divisions, representing the half-hours in a twenty-four hour day. In the centre of the board are placed stacks of cardboard action counters, representing the various types of action which might be made by a commander during the day. Each counter should be cut to size so that it can fill the appropriate number of time divisions on the board; i.e., a counter for a half-hour action would be the size of one division; for a one-hour action it would fill two divisions (for the precise list of actions see the rules, below). During play the counters are distributed round the game board in each twenty-four hour turn, so that every division is filled by one counter. This shows how that commander intends to use his time.

In addition to maps and game boards, players and umpires should all have plenty of spare paper and summaries of the rules. Umpires should also have a checklist of factors which will need attention, and one nugget.

Playing the game

The umpires must set a problem, either from imagination, or from the history of Napoleonic campaigns. This will require a schematic map to be drawn of the theatre of war, an order of battle, i.e., a list of the army corps, but nothing lower than that, and a starting date for the action. If we take the Jena campaign (1806) as an example, we might find that the French base is the fortress of Mainz; the Prussians' is Berlin; and that both sides have five line corps, one reserve corps, and one cavalry corps. The start of the game is set at 1 October, when the French are echeloned between Mainz and Bamberg; the Prussians between Berlin and Erfurt.

The umpires must next decide the status score of each corps. For simplicity we will say that every line corps starts with 25 points; every reserve (or Guard) corps with 40; and every cavalry corps with 10. The commander's personal staff, the little HQ and the grand HQ have no status allocated to them.

The commanders-in-chief must also have allocated to them the number of hours' rest they will ideally need every day. For the young Napoleon, for example, we may allocate a requirement of nine hours out of every twenty-four; while for the Prussian Duke of Brunswick, at seventy-one years of age, we will set the figure at twelve hours.

Finally, the umpires must state which towns are fortresses, which are open supply points, and what supply each contains. Fortresses normally contain enough supplies for one corps to live on indefinitely, plus fourteen corps days of additional supply, i.e., they may supply a second corps for fourteen days, or seven extra corps for two days. Normal supply points can provide for one corps indefinitely, plus seven extra corps days. Large towns and exceptionally rich areas, however, may be rated much higher. In this case Berlin, Mainz, Leipzig, and Erfurt should be given a rating of twenty corps days each, in addition to supplying one corps indefinitely. The supply for the first corps may not be split between others, i.e., a town which has exhausted all its additional supply may not re-allocate its one corps ration, indefinitely, in the form of an indefinite number of rations on one day.

Sequence of actions in each turn

To begin, players lay out their unit counters in the starting positions on their maps, and commence action according to the following sequence for each turn, which represents twenty-four hours, starting at midnight:

(*i*) Umpires give players any intelligence of the enemy so far received.

(*ii*) Each player arranges a selection of action counters around the perimeter of his game board to fill up all of the forty-eight half-hour divisions. This shows exactly how each half-hour is to be spent by the commander.

(*iii*) Players may now use a separate piece of paper to make additional notes about certain action counters which they may have used.

If any orders are to be sent to mobile units, the content of the orders is noted, together with the times the messages would arrive. Remember that the three types of HQ count as mobile units in this context.

If any intelligence is to be gathered, the specific towns to be investigated must be listed.

If any inspections are to be conducted, the specific unit or supply point concerned must be listed.

(*iv*) Umpires collect all this information from the players, and work out the movements of both sides on the master map. If any opposing forces have actually come into close combat with one another, as opposed to cavalry screen contact which does not imply combat itself, the timing of the combat is noted. The distance between the combat and the commander-in-chief is also noted, and a courier is deemed to have made a combat report to the commander-in-chief as soon as he could have covered that distance, e.g., if a combat starts at Erfurt at 1200 hours, and Napoleon is at Gotha, one march away, then the report will arrive after the time needed for the courier to travel one march, which is four hours. In this case Napoleon would be told of the combat at 1600 hours.

(*v*) If there has been a combat report, players now have the option of rearranging all their action counters for the rest of the day, after the arrival of the report; in this case Napoleon might want to drop what he had intended to do after 1600 hours, and move quickly to the battlefield instead.

(*vi*) Umpires now take full note of all orders finally issued by players, and assess the results of that turn. This includes:

The results of combats are found, and movements adjusted accordingly.

Adjustments to the status of units are noted, from whatever cause.

Adjustments to the supply rating of all towns are noted, plus any servicing of the line of communication which has been left outstanding.

Any rest or letters home which have been left outstanding are noted.

(*vii*) Umpires now report back to each player any of the above results which would be known to them as at midnight.

Intelligence reports also now arrive from all screen contacts (cavalry scouting one march ahead of every corps) and from any intelligence investigations which have been ordered.

(*viii*) Players bring their maps up to date, clear their game boards, and start preparing their actions for the next turn.

Actions open to commanders in each turn

Players may choose combinations of the following actions for their commanders to make in each turn, to a total of twenty-four hours' worth. Action counters will then be laid out in the relevant divisions of the game board to represent each type of action.

Writing orders for mobile units

An order for one subunit takes a total of half an hour to write and dispatch, thus a movement order for all nine units would take four and a half hours to complete. A short cut may, however, be made by ordering movements several days ahead for the same unit, i.e., only one message would need to be written for that unit during all of those days.

The player must work out the sequence in which he wants the orders to be written, and the times it would take for each to arrive at its destination. Thus during a session of writing which started at midnight, five letters might be composed at half-hourly intervals. The first, ready at 0030 hours, might be for a unit one march away. It would take the courier four hours to cover the march, so the letter would arrive at 0430 hours. The second letter, ready at 0100 might be for a corps three marches away, which would take the courier fourteen hours to reach. This letter would therefore arrive only at 1500 hours.

Writing letters to service the line of communication

As the army advances, it must establish a line of communication, i.e., a continuous line of towns which can be traced from the army's base to all its units. The line must then be maintained administratively, as well as protected from enemy attack. The base itself will always require one letter per day, and every multiple of four towns in the line will also require one letter. Thus if the line consisted of the base plus nine towns, that would require four letters per day (odd numbers are always rounded up). These letters, as for movement orders, will always require half an hour each to write, i.e., a total of two hours in this case.

Establishing a town as part of the line takes no extra time, as soon as a friendly corps or HQ has passed through; but changing the line, establishing a new base, or constituting a new HQ if any HQ has been overrun by the enemy will require four written orders and take two hours of the general's time. Note especially that all correspondence for the line of communication must be conducted in the presence of the little HQ.

If the commander fails to service his line of communication fully in the course of any day, or if the enemy breaks through it, there will be serious effects upon the army's status.

Intelligence

Every hour a player allocates to intelligence gathering during the day entitles him to a spy's report on one town anywhere on the map at the end of that day. The report will give an accurate list of all units passing that town during the day, but not their current status. Note that all correspondence connected with intelligence must also be conducted when the general is with the little HQ.

Writing home

Writing all the assorted letters home for the commander's family, government and allied governments will take one hour per day, or two hours in two days, in the presence of the little HQ. If it is delayed more than two days, however, commanders will be forced to spend the first four hours of the third day making up for lost time.

Any correspondence destined for other players, such as a call for the enemy to surrender, or a suggested plan for an allied field commander, will take a whole hour per letter. This applies even when two allied players are supposed to be at the same place on the map, they must still communicate in writing, since the two players will not be allowed to chat to each other directly.

Note that all letter writing, for whatever purpose, must be conducted when the commander is stationary: writing will not be allowed during travelling time.

Rest

Rest may be taken at a halt, or up to half of it may be taken during movement in a coach (as opposed to movement by post-chaise or on horseback).

If the commander fails to take enough rest during the day, the excess is carried over into the next day's requirement. A general who consistently goes short of rest will therefore build up a backlog of inertia. When the backlog reaches twenty-four hours it must all be taken at once.

Note that extra rest will be required after horse riding, or if a rest period is cut short unexpectedly by a battle or battle report.

Wellington conducting a hasty inspection to inspire his troops before battle. Note that although this could be performed on horseback, the sheer number of troops to be inspected would make it rather a lengthy process.

Movement around the theatre of war

The commander may move by horseback without reliefs, unless an order to prepare posts has been sent to the little HQ the day before; by post-chaise; or by coach. Sustained travel is possible only in a coach or on horseback with posted reliefs; but any type of horseback movement requires extra rest. Note that the commander should not abandon his little HQ for long, as he will require it for servicing the line of communication, for intelligence, and for writing home.

The grand HQ, on the other hand, need never be visited by the commander. It must always be within four marches of the reserve (or Guard) corps, otherwise that corps will lose status, because, for example, the reserve artillery lacks administrative support.

Inspection of troops and supply points

When a commander inspects a unit he can raise morale and ensure that orders are being carried out. In the case of a town in the line of communication, inspections can spur the local commissaries to collect more supplies from the surrounding country. We distinguish two types of inspection: hasty, lasting half an hour, and adding one status point to a corps, or one corps day extra supply to a town; and formal, lasting four hours for the inspector, but immobilizing the inspected unit for the entire day. This raises the corps' status by four points, or adds four corps days of extra supplies to a town.

No corps or town will be allowed more than one hasty and one formal inspection per week, to prevent the stockpiling of unrealistically huge status or supply scores.

Speeches before combat

If the commander-in-chief can arrive at a battlefield before the battle actually starts, he may harangue each corps in turn. Each harangue will raise that corps by two status points, and will take half an hour. Only one speech of this type may be made to each corps each week.

Personal reconnaissance

Once again, this depends upon the commander reaching the field of battle before battle is actually joined. Reconnaissance may be done on foot, taking four hours and adding three status points to every corps in the battle; on horseback, taking two hours and adding two points; or in a post-chaise, which takes two hours and adds one point to each unit.

Whenever a commander conducts a personal reconnaissance, a nugget is rolled. He is killed if it comes up 0, or wounded for a 1 or 2.

Battle

When two opposing forces run into each other during a move, the sequence for combat is followed (see below) and a battle may result. If the start of the battle is delayed for any reason, the commander-in-chief may have an opportunity to conduct a personal reconnaissance, or to harangue his units. When a battle does finally start, if a commander-in-chief is present he must choose his own position in each of the three phases of combat. In each phase he may opt to fight in the front line, giving personal direction, but risking being hit; or he may be safely behind the action, enjoying a more balanced view of the whole. In the former case a new nugget is rolled in each phase of combat to see if he is hit; 0 he is killed, 1 or 2, and he is wounded. He also adds two status points to each of three corps, but is debarred from writing strategic orders to units outside the battle area.

If the general opts to stay behind the lines during the battle, he is in no personal danger, does not add any additional status to his units, but may continue all types of letter writing in the normal way, at the same time as commanding the battle (which will not take him any extra time).

If the commander-in-chief is totally absent from the battle, or if he is hit during it, the battle is fought under the absent general rule (see below). If the commander is hit, all units in the battle lose a status point, and a hand-over to his second-in-command takes place. The hand-over takes one whole phase of the battle to complete.

Movements by corps in each turn

Units take the following number of hours to move from one town to the next along a road, i.e., to make one march. Forced marches are additional to the normal marches taken during that day.

Type of unit	*Normal march*	*Forced march*	*Comments*
Cavalry corps	6	6: one every day if desired	Lose 1 status point for every forced march
Reserve or line infantry corps	8 or dig in	8: twice a week maximum	Lose 1 point for each march or forced march.
Grand HQ	8	–	–
Little HQ	4	4: once a day	–
Courier	4	5: as often as desired	i.e., the courier covers the first march in 4 hours, and each march after that in 5 hours. May move cross country if desired.
General and personal staff			
Horseback without reliefs:	3	3: once per day maximum	Requires one extra hour's rest per march/forced march.
Horseback with reliefs pre-posted:	3	3: as often as desired	As above
Coach	4	4: as often as desired	May sleep *en route*.
Post-chaise	3	–	–

If horses are used, the commander may leave the roads and move cross country.

Combat

When a force comes within one march of an enemy force, it automatically receives news, at once, from its own cavalry scouts. It is therefore impossible for a unit to approach nearer than one march to an enemy without being reported.

If the two forces then decide to continue towards each other, and run into each other on the same spot, the combat sequence is started.

For convenience the combat sequence is assumed to happen at the nearest town. No measuring of distances along roads is necessary. As soon as the combat sequence starts, couriers are also assumed to be sent immediately to the two commanders-in-chief.

Both sides then state whether they wish to attack, stand, or withdraw, as well as which corps, if any, they are keeping in reserve, out of the initial contact.

If one side attacks while the other either stands or attacks, there will be an immediate battle. If one withdraws while the other attacks, then there will be a two-hour pause for manoeuvre followed by a new statement by each side whether it wishes to attack, stand or withdraw. Provided that the original attacker still opts to attack, then there will automatically be a battle, i.e., the withdrawing force may be able to delay the combat, but cannot avoid it for more than two hours.

If neither side attacks, there will be no immediate battle. One side may wish to take advantage of this to withdraw, or both sides may stand watching each other, and perhaps also digging in. Either side may change the orders and attack or withdraw before battle starts.

The commander-in-chief may already be present; he may arrive during the battle; or in the lull beforehand. If he finds time before the battle, he may make a personal reconnaissance, or harangue the troops.

No fighting may take place at night. If night falls (usually between 2000 hours and 0400 hours, although in winter it may be 1600 to 0800 hours) during a battle, the battle will be interrupted at the end of that phase, and one side or the other may disengage if desired. Generals may always make their reconnaissances or speeches at night.

The battle is divided into three phases, each of which lasts as many hours as there are corps on the weaker side at the start of that phase. Reinforcements may not be fed into a battle in the middle of a phase, but always at the start of the next phase, if they are available.

At the start of each phase the commander-in-chief, if present, must declare whether he will fight in the front line, or behind the battle. Whichever he decides, he must then also declare whether his troops are attacking, standing or withdrawing during that phase, and whether or not he is committing any reserve units to the battle in that phase. If the general is absent, or has been hit, a nugget is thrown. The decision for that phase will not be executed for a 0 or 1 (the absent general rule).

(1) screen contact

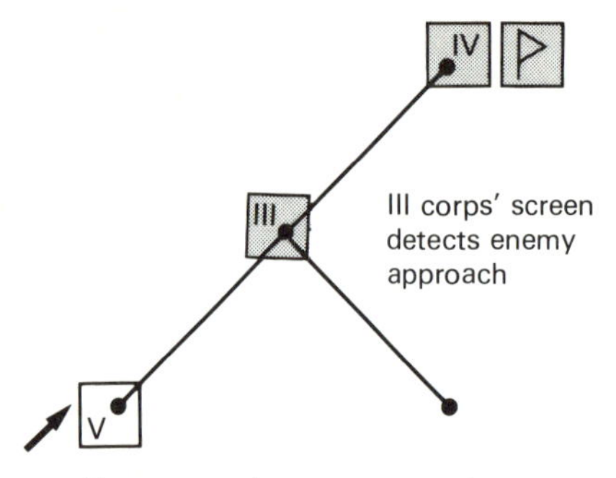

V corps marches to one march distance from III corps – receives screen report

(2) start of combat

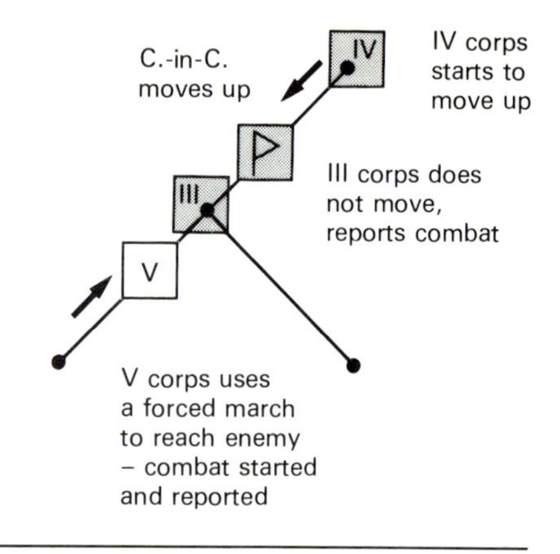

(3) phase 1

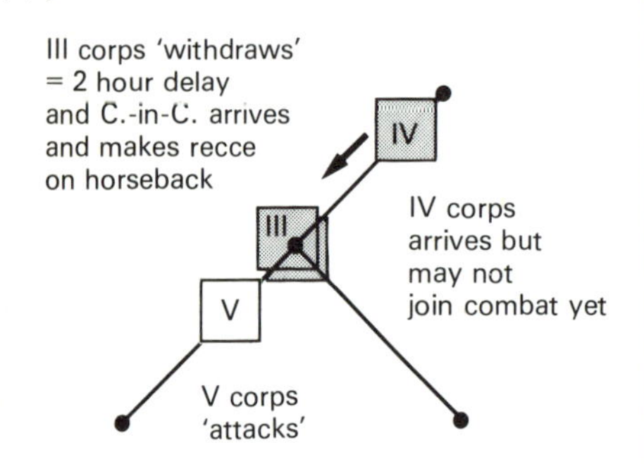

(4) end of phase 1

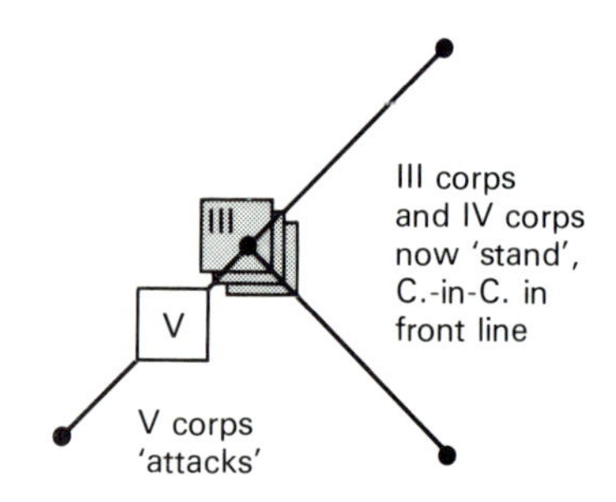

The result of phase 1 is assessed. V corps is the loser, and the enemy C.-in-C. is not hit in the front line.

(5) phase 2

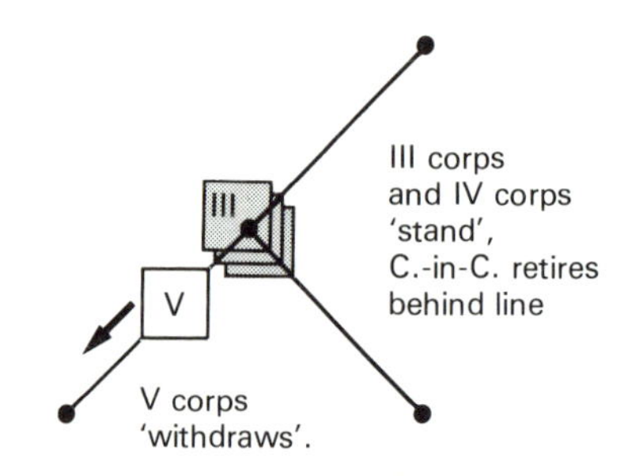

V corps has successfully disengaged before the next phase is assessed, but has no movement left to spend.

(6) new combat, phase 1

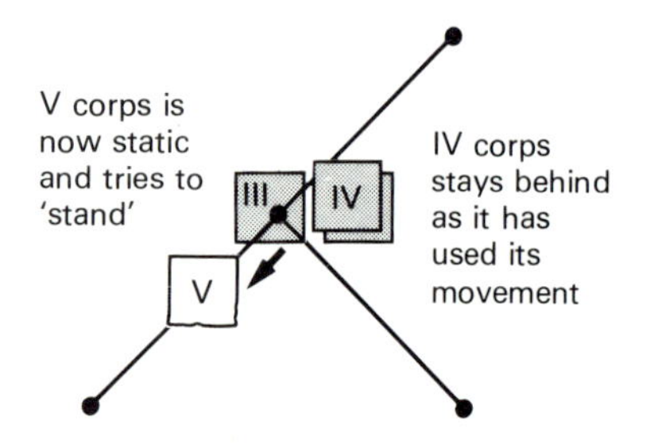

Seeing V corps has run out of movement, III corps immediately moves up, which starts a new battle. III corps attacks in phase 1. The result is now assessed, and so it goes on.

An example of the combat sequence

After the decisions have been taken for each phase, the total status of all committed units is found, excluding any troops who have arrived at the scene of battle, but who are being held in reserve. All status will now be brought up to date, including the result of any personal reconnaissances, commanders fighting in the front line and so on.

Any unit which is standing behind completed fieldworks may add a quarter of its score again. If a unit is fighting beside a friendly fortress, it adds half its score again. Distinguish this type of battle from an attack on the fortress itself, which counts as a siege (see below). If one side has a cavalry corps when the enemy has none, the cavalry corps doubles its score, provided that there is also friendly infantry on the battlefield.

When both sides have found their total scores, after adjustment, one nugget is thrown for each side. For 0–5, that side's score remains unchanged. For 6–8 that side multiplies its score by 150%. For a 9, the score is doubled. The higher final score wins that phase and loses one tenth of all status scores originally committed to the action. The loser loses one fifth of all status scores originally committed to the action. If a withdrawing force wins a phase, it may disengage. Otherwise it must continue to fight, unless it has a cavalry corps while the enemy has none, or can retire directly into an adjacent fortress.

After the result of the phase has been found, the next phase is played. This continues until the battle ends after three phases, at nightfall, or when one side or the other successfully disengages.

Sieges

When a fortress is attacked, a field force may be present. In this case there may be a normal battle, or the field force may opt to take shelter inside the fortress. It may retire in this way whenever it wishes, and may remain in the fortress for as long as the supply lasts out; one corps could remain indefinitely, but a second one would normally have supplies for only fourteen days.

If a fortress is attacked when there are no field units present, or after field units have retired into the fortress, the attacker may either pass through the town unhindered, or he may blockade or besiege the place, in which case supply becomes critical. One attacking corps will automatically find enough supply to continue indefinitely; but additional units will be forced to disperse to surrounding towns to find supply. The garrison of the fortress must also establish how many rations are available. This is the number of days shown by one nugget roll multiplied by ten, e.g., a roll of 6 would mean the garrison could hold out for sixty days. If there is a blockade, the blockading force surrounds the fortress until the fortress runs out of supply. The garrison, however, may attempt a sally at any time, which is fought as a normal battle.

(1) investing the fortress

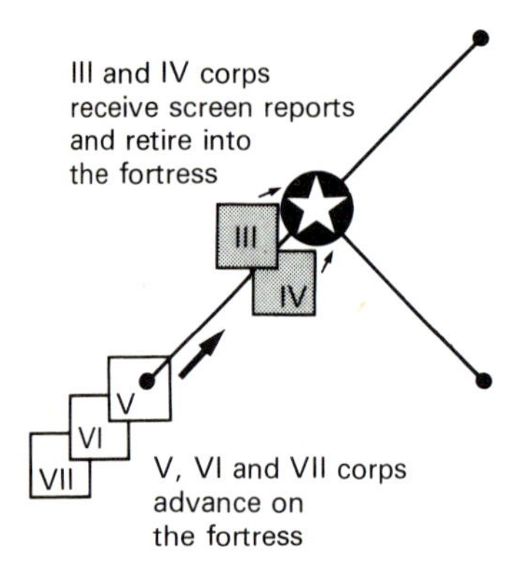

(2) completing investment

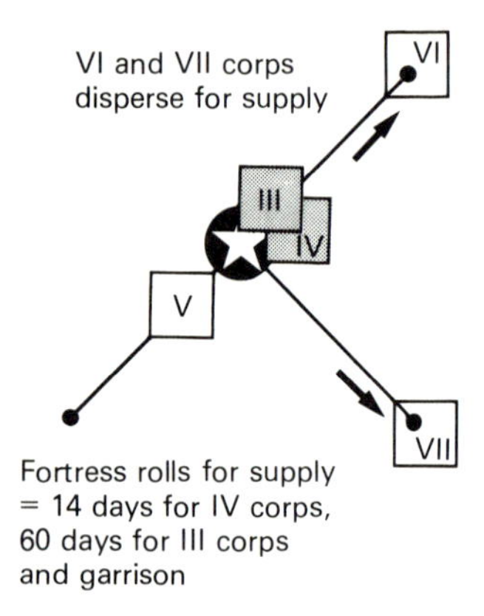

(3) approaches

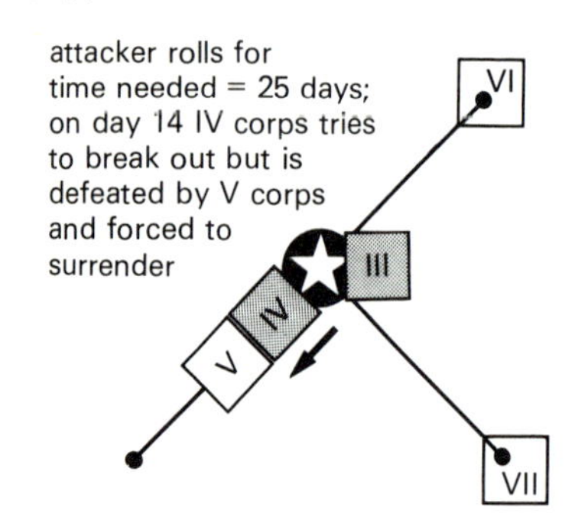

(4) breaching

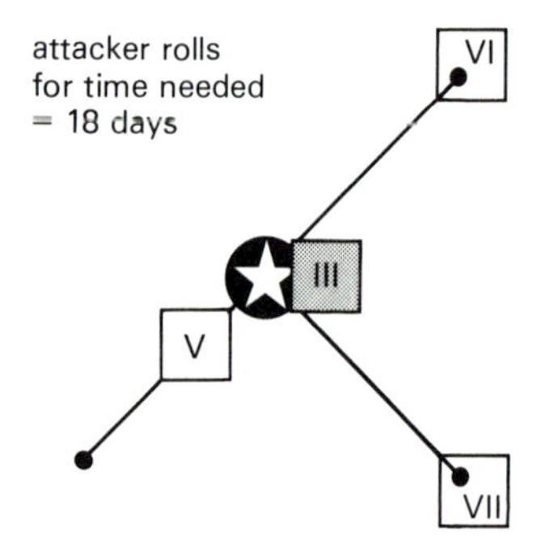

(5) 1st storm

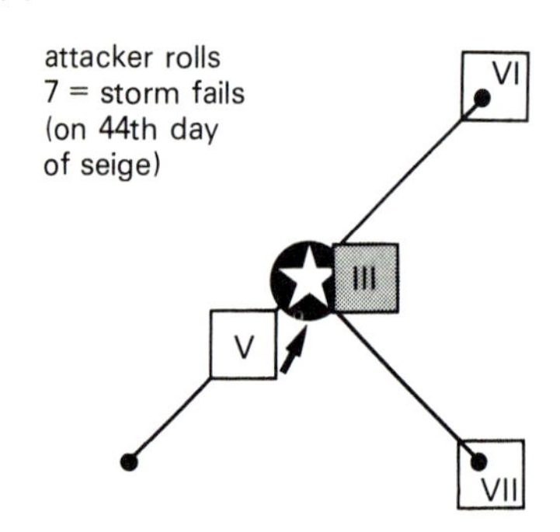

(6) 2nd storm

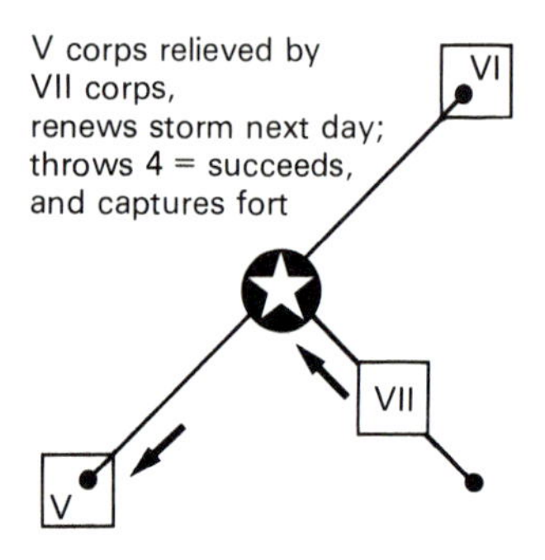

Total loss to attacker = 5+6+7+4 points (scores on dice rolled in the three phases)

An example of the seige sequence

If the attacking force makes a siege, they advance in three stages:

The approaches This phase lasts the number of days shown on one nugget multiplied by 5, but if a 0 is rolled, the siege must be called off.

Breaching This lasts the number of days shown on one nugget multiplied by three, but once again, a roll of 0 means the siege must be called off.

Storming This takes one day, and succeeds for a roll of 5 or less on one nugget. If it fails, it may be renewed next day.

Every single point shown in each of these three nugget rolls by the attacker represents one status point lost from his total score. If the siege fails, the defending troops lose the number of points shown on one nugget roll, but may instantly repair the fortress so that any new besieger will have to start from scratch. If the siege succeeds, on the other hand, all defenders are captured.

Alterations to status

Units lose status, as already indicated, after each phase of a combat or siege, or after every forced march they make. Infantry line and reserve corps also lose one point for every normal march made. All units in a battle lose one point immediately if the commander is hit. Any unit which arrives at the end of the day at a town which cannot supply them will also lose one point, and any unit unable to trace a continuous line of communication to the base will lose a further point for every day this continues. The same is true for all units if the commander has failed to service the line of communication adequately in that day. The reserve corps loses one point for every day grand HQ is more than four marches away; but all units lose three points if grand HQ is overrun by the enemy. One point is lost by all units if little HQ is overrun.

Units gain status, as already described, if the commander fights in the front line, makes a personal reconnaissance, harangues the unit, or inspects it. Units also add one status point to their score up to, but not beyond, their original starting score for every day they remain in the same place, provided they do not run out of supplies.

7
Map kriegspiel

The generalship game is almost a boardgame, and relies upon a lengthy set of formal rules. It undoubtedly forces the player to concentrate very hard upon what a Napoleonic general would have been doing with his time; but due to its complexity it may not be to everyone's taste. As a counter-balance, therefore, we now turn to a game which has almost no rules at all. This is the free kriegspiel, played on maps.

Origins and advantages of kriegspiel

Early recreational wargames were rather abstract and unrealistic affairs, usually based upon some variant of chess. During the nineteenth century, however, wargames tended to develop a more realistic format, largely as a result of the renewed military interest caused by the Napoleonic Wars themselves. Thinking officers were then starting to experiment with various ways of representing battles in miniature; and in Prussia this developed to a very advanced stage indeed. The military kriegspiel gradually became a recognized means of officer training, and later even evolved into an aid for strategic planning.

The nineteenth-century Prussian game started life with a rigid structure and copious formal rules. The two sides were each placed in a separate room with a model of the terrain or a map. The umpires moved from one room to another collecting orders from the players, and then retired to a third room to consult the rules and find the results of combat. A great deal of their time was consumed in leafing through voluminous sets of rules, consulting tables and giving rulings on fine legal points. By about 1870, however, this rigid system was starting to be thought rather clumsy and time-consuming. Quite apart from the many defects and loopholes in the rules themselves, it reduced the umpires, who were often very senior officers, to the role of mere clerks and office boys. Clearly, such a state of affairs was intolerable.

It was General von Verdy du Vernois who finally broke with this system, and abolished the rule book altogether. His approach to the wargame was the free kriegspiel, in which the umpire had a totally free hand to decide the result of moves and combats. He did not do this according to any set of written rules, but just on his own military knowledge and experience. He would collect the players' moves in exactly the same way as before; but he would then simply give a considered professional opinion on the outcome. This speeded up the game a very great deal, and ensured that there was always a well thought-out reason for everything that happened. This was a great help in the debrief after the game, and it allowed players to learn by their mistakes very quickly.

The free kriegspiel using maps can offer many advantages for modern wargamers provided that the umpire has a reasonable background in wargaming, and a bit of common sense. If this condition is met, the game immediately becomes faster and less pedantic than if it had been tied down to a set of rules. The umpire can always think of more factors to incorporate in his decisions than could ever be true in a formal or rigid game. He can therefore spread a greater atmosphere of realism about the game.

What you will need for a map kriegspiel

The umpire must be someone who knows as much about the Napoleonic Wars as the other players, so that he will be able to keep a little ahead of their criticisms. In fact this superior knowledge need not amount to a very great deal, and even relative beginners will be surprised at how easy it is to umpire a game of this sort. They should not be put off by the fear that umpiring needs some formidably experienced military brain, like that of General von Verdy du Vernois himself: it doesn't. Almost anyone can do it, with a little practice. Apart from anything else, the umpire always has the advantage that he is the only one who can see the complete picture of what is going on.

It is best to have three rooms, one for each team, and one for the umpire; but at a pinch the umpire can do without his, and simply keep moving from one side to the other, making notes behind the backs of the players. This also economizes on maps, as the umpire will not need one. For very elaborate games, on the other hand, any number of rooms may be used, and the author has participated in some games using six different playing teams, as well as a sizeable team of umpires.

The maps themselves may pose a problem, since they can become rather expensive if bought in bulk. You should therefore choose the particular game you are going to play rather carefully, with this in mind. Clearly it depends a great deal on your financial circumstances; but it is perhaps worth reflecting that a set of three Ordnance Survey maps will cost rather less than the average boxed boardgame. At any event, if all else fails you can always make your own sketch maps of the area to be fought over, with traced copies for all the players. For sieges and some tactical actions, indeed, this method will be the only one possible.

You may wish to mark movements on the map with a set of pins, but it is usually easier, and better for the map, to use a talc overlay and a set of chinagraph pencils. In this way movements can be shown graphically, explanations pencilled in, and the whole thing will be easier to understand.

Players and umpires will require rulers, plus carbons and spare paper for writing reports and notes. The umpire will also require one nugget.

Playing the game

The umpire will select a scenario which fits onto the available maps. One hardy perennial (which uses the British 1:50,000 Ordnance Survey series) is a hypothetical landing by a French corps in some part of the British Isles. The French are allowed to land *en masse*, whereas the British troops start the game widely scattered. Political aspects may also be incorporated into this game, with Jacobite sympathizers and other adventurers fighting their own guerrilla wars in the back hills.

If you use continental road maps it is perfectly possible to re-fight all the classic operations of Napoleon. Admittedly the maps will be quite small scale (1:500,000 or thereabouts); but then so were the maps Napoleon himself had to use. If one is operating with a number of army corps the large scale details of the terrain will not be important, in any case.

Another alternative is to fight a siege. For this you can either photocopy the plan of a real fortress, and use that for your map; or you can draw your own fortress plan from scratch. One player is the defender, and must move his batteries about inside the fortress, perhaps digging mines under the attackers' trenches, and occasionally making brief sallies. The other player will have to dig trenches up to the fortress, so he can establish breaching batteries and eventually storm the breach. A free kriegspiel is particularly suitable for playing siege operations, since the tedious repetition of many siege operations can be rushed through by the umpire to fit the available time. They do not have to be played through in minute and boring detail, as they would in a game with rigid formal rules.

Order of battle

When the umpire has selected his map and set a problem for both sides, he must give all players a full list of their forces, and keep a carbon copy for his own reference. Note that the umpire will give information to players only about their own forces, with very few clues about the enemy's. Players will then be fed snippets of intelligence about the enemy according to the types of reconnaissance they ask for. They will have to build up a picture of what the enemy is doing for themselves from this information.

The umpire finally states the date, time and weather at the start of the game.

Sequence of turns

The game progresses in a series of turns, in each of which the following sequence is observed:

(*i*) Players write orders and pass them to the umpire.

(*ii*) The umpire compares the orders from each side and decides what sightings and contacts have been made, and at what times.

(*iii*) The umpire may then wish to ask players for supplementary information; e.g., if there has been a contact between two opposing formations, the umpire may need to know whether players want to withdraw, or to stand and fight.

(*iv*) The umpire then decides the result of combats, and the reports to be given to players from combats and other sightings.

(*v*) The umpire reports all this information to players, who start writing orders for their next turn.

Each turn will usually represent twenty-four hours of the campaign, as in the generalship game. This allows a convenient cycle of actions to be completed, and is realistic in the sense that Napoleonic commanders did tend to write their orders at the same time each night. If a particularly large order of battle is being used, however, such as a large number of army corps, then a two- or three-day cycle may be preferred. If only small units are being used, on the other hand, it may be better to use a three hour or a six hour cycle.

With a little bit of experience umpires may be able to break away from a regular cycle of turns altogether, and start to tailor each turn to the tactical needs of the moment. Thus if not much is happening in the game, for example, during the lengthy digging phases of a siege, several days may be covered in a single turn; whereas if the action is fast and furious, say, at the moment when a breach is stormed, only an hour or two will be covered. The umpire must decide roughly how much time would have elapsed in real life before the players would have had to make each important decision. The turn will then be extended or contracted so that it represents that amount of time. Each turn, in other words, should include one moment of decision for each of the players.

Movements

The umpire, as in all aspects of this game, has the last word on how far or fast units have moved. For the guidance of players, however, a rough sheet of planning figures ought to be provided, something like this:

KILOMETRES MOVED DURING THE AVERAGE DAY

Type of troops	*Km moved*	*Comments*
Infantry	21	–
Arty & vehicles	21	Must stick to roads, delayed by bad weather.
Heavy cavalry	25	–
Light cavalry	28	–
HQ group	31	Move any time of day or night.
Couriers	6.5 km per hour for first 4 hours = 26 km, then go 4 km per hour after that, indefinitely.	

The umpire should also keep certain brief notes for his own guidance, e.g., the couriers may fail to arrive if a nugget comes up 0; or the ratings of rivers and bridges may be decided in advance, so that players who send out scouts to look at such matters may be given a clear answer, and so on. The degree to which notes of this sort are made will depend a great deal upon the individual umpire. In many cases rulings can be made *ad hoc*, as and when they are required.

If the game is to be a siege, a similar table of moves and timings may be kept for the actions appropriate to siege warfare:

DIGGING POSSIBLE DURING AN AVERAGE DAY

In twenty-four hours each working party may:
Dig about 70 m of sap
Build one third of a battery
Build half of an infantry redoubt
Dig 5 m of mine gallery
Arm a battery, i.e., put in cannons
Arm a mine, i.e., put in a charge of powder

Once again, the umpire will use these figures as a rough guide, and alter them according to the various changing circumstances; in bad weather or under heavy enemy counter-fire, digging would be slowed down.

Combat

The system for finding the results of combat in a free kriegspiel is classically simple. First of all the umpire looks at the position of each side: how many and what type of troops are involved; how their morale is bearing up; and what orders they have been given. He next considers the ground on which the action will be fought, and any special tactical problems which either side might encounter; whether there are any obstacles in the way of an attacker; whether a flank attack might be possible, and so on.

When the umpire has all relevant information at his disposal, he ought to be able to give an informed opinion on the probabilities of the result. He will not simply say something like 'The French infantry has

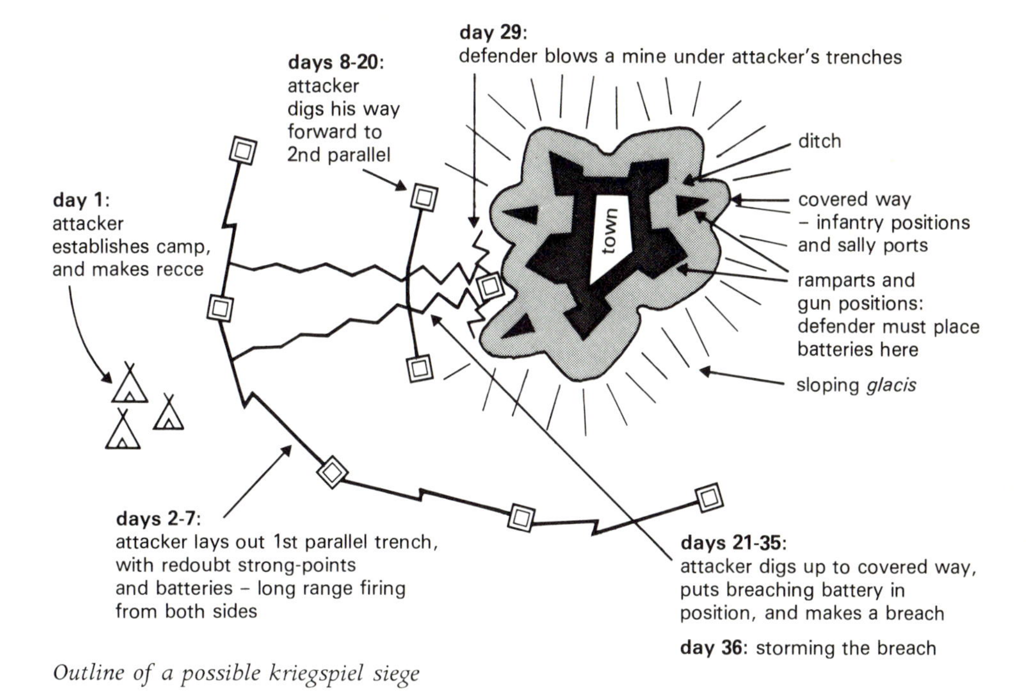

Outline of a possible kriegspiel siege

successfully stormed the hill', but will quote possibilities, such as: 'The French have a 50% chance of storming the hill successfully; a 30% chance of capturing half of it, while disputing the rest; and a 20% chance of being totally repulsed. High scores favour the French'. It is important that the umpire is as specific as possible with these figures, as this forces him to consider all the factors involved in the combat and to think through the full implications of his decision. He must also be clear whether a high dice roll will be good or bad for the attacker, i.e., whether the top 50% (a die roll of 5–9) or the bottom 50% (a roll of 0–4) will mean the hill has been carried. In this case he has stated that the high score will be good for the attacker.

Finally, after odds have been quoted the umpire rolls a nugget, to represent chance. This will give a percentage, from which the final result of the combat may be read off. Thus in our example a nugget score of 1 would be under 20%, so the attack would be repulsed. A score of 8 would be within the top 50%, so the attack would succeed, and so on. The system works by the umpire giving his opinion on the probabilities, and then rolling a nugget to find which of the possible results actually came up.

Let's take another example, from siege warfare. The fortress may be firing at a particular trench with four cannons for twenty-four hours. The umpire will see what size of guns are firing, and what the diggers are up to. He will then assess the terrain, and find whether enfilade fire is possible. He may then give his opinion that there is a 10% chance of digging being halted by the fire with 100 casualties; a 40% chance of digging being slowed down to half-rate with 60 casualties; and a 50% chance of it going at three-quarter rate with 40 casualties. He announces that high scores will favour the fortress, and rolls a nugget. If it comes up 3 he knows that it falls within the bottom 50%, so digging goes at three quarter rate, with 40 casualties. Had the score been 9 it would have been in the top 10%, so digging would have been halted for that day. Remember that in all this the umpire has to be certain of what each nugget score will mean, before it is thrown.

These are all the rules required for free kriegspiel. It is a remarkably straightforward game; but it can produce some of the best results of all. It allows speedy resolution of combat; yet at the same time screens the players from any unrealistically panoramic views of the battlefield. All it needs is someone who will not be overawed by the responsibilities of umpiring.

8
Tactical exercise without troops

In our seventh and last Napoleonic wargame, we approach the problem of realism from a radically different direction. Instead of stressing tactical manoeuvres (as in the games with model soldiers) or the organizational activities of high commanders (as in the generalship game), we now concentrate upon the battlefield itself, the ground over which generalship must be exercised.

Origins and advantages of the tactical exercise without troops

In the Napoleonic period it was recognized that a successful commander needed what was known as *coup d'oeil*. This meant that he had to be able to assess the military possibilities of ground, both quickly and accurately. He had to relate abstract map manoeuvres to a real piece of terrain, and choose his positions accordingly. This skill was fundamental to everything which happened in battle, and the commander with the better *coup d'oeil* would usually enjoy a marked advantage over his opponent.

Because of this, Napoleonic officers were often encouraged to study the art of field sketching, and even surveying. Staff officers would also be expected to make full topographical reports on the terrain through which they had passed. These reports often attained a higher quality than the printed maps available at the time, and generals were frequently forced to rely upon them while making their strategic decisions.

During the nineteenth century this straightforward and purely topographical training was supplemented by the addition of truly tactical problems. Officers were taken out to a piece of real ground and tested in their ability to plan troop deployments upon it. At first this type of exercise was called a staff ride; but as time went on it was renamed a tactical exercise without troops (TEWT) and it has now been a standard part of officer training for over a century. Its attraction is that it uses real terrain rather than simply a map, so it can give a much clearer idea of exactly what tactical possibilities are open. A 25% slope, for example, may not be very revealing as an abstract concept but when you actually have to climb up one its full meaning may suddenly become very obvious. To take another example, it may appear possible, from the map, to establish a battery at a particular position to give supporting flank fire for another part of your line. On the actual ground, however, there may be unexpected problems such as scrub obscuring the view, or boggy ground on which guns cannot be deployed. In these cases the battery commander may have to think again very seriously about his tactics.

In recreational wargames it is very easy to forget the realities of terrain and depend entirely upon the map, although that is actually no more than a form of shorthand note. In many wargames the map is even deliberately simplified to assist the mechanics of play. This habit may well bring several advantages to the game; but it certainly also reduces the wargamer's need for a genuine *coup d'oeil*. If we want to be realistic in our representation of Napoleonic command, however, we must give *coup d'oeil* a much more important place of honour in our games.

We have two approaches to choose from. The first is simply to get to know the countryside, try to look at it from a staff officer's point of view, and then incorporate our findings into normal indoor wargames. This means that we should examine the military possibilities of the landscape whenever we go on country rambles and think about such things as fields of fire, or the firmness of the ground for cross country movement. We should try to estimate how long it would take a battalion to cross each river, or how difficult it would be to pass artillery through each wood. Last, but by no means least, it was always important for Napoleonic staff officers to know the resources of the country; how much wood and water were available for bivouacking, and how much food and forage. We should therefore practise estimating the subsistence available in each area we pass.

Linked to this approach, we may sometimes be able to visit genuine Napoleonic sites, and try to relate the lie of the land to the tactical problems. This, unfortunately, means that we must travel to the continent to find true battlefields; although in Britain there are admittedly a few fortresses of about the right vintage. By inspecting them we can at least learn a bit about the ideas of Napoleonic gunners and engineers.

Our second way of working *coup d'oeil* into wargames is to make an adaptation of the TEWT. We can convert the one-sided military version, where the student plays only against the umpire or instructor, into a two-sided, competitive wargame.

What you will need for a tactical exercise without troops

The most essential thing for this type of game is that all participants must be sympathetic to the aim of developing *coup d'oeil*, and will not be too easily discouraged if early attempts do not work very well. It is useless to drag people into the countryside if they don't understand, or don't like, the reasoning behind the TEWT. This game, in other words, is not for beginners. It requires more co-operation and mutual agreement than almost any other wargame, and has the least room for cheating and gamesmanship. Despite its deceptive simplicity, it requires a great deal of careful thought. If there is an umpire, he must be particularly imaginative and attentive to detail.

The number of players in a TEWT may vary from two to half a dozen or so. The use of an umpire is optional, although it becomes more advisable with more than two or three players.

TEWT wargames may be either pre-planned or spontaneous; but the more people are involved, the more necessary it becomes to make a careful plan in advance of the actual game. Pre-planning offers many advantages for the game mechanics; and players will start with a good idea of what it is all about. The spontaneous TEWT, on the other hand, will tend to fall flat more often, but it is also less formal, and can sometimes produce the most memorable and natural games of all.

For a pre-planned TEWT the umpire, or the two opposing commanders, must first decide on an area of countryside for the game. They must reconnoitre it for access and visibility, and fix the size and aims of the two opposing armies. The type of combat to be presented must be envisaged, and so must the particular problems of the ground, and the possible complications which may arise. Only after this sort of detailed preparation has been done can one start to bet that the TEWT will be a success.

The spontaneous TEWT usually arises when two wargamers happen to be walking together in the countryside, or when they are at a loose end near a piece of countryside which they know to be suitable for wargames. One of them will probably say something like, 'What would you do if I had a battalion and two field guns in that farm, and you had a regiment of infantry to get me out?', or 'I will deploy my Division along the reverse slope of that ridge: what are you going to do about it, if you have only another Division and a brigade of cavalry?' Another variant, which fits particularly well into a long walk, is to assume that one side is marching a convoy along the road you are using, while the other side is constantly trying to ambush or harass them with a few guerrillas or irregulars, with perhaps some regular riflemen in support. This type of game can go on for miles and miles, as the marching unit plugs on despite its casualties, inflicting more or less damage to its assailants.

You must find a suitable piece of countryside for a TEWT. The first requirement is that it should be reasonably Napoleonic in appearance, with only a few buildings, preferably of the period in design. A sprinkling of thatched cottages is fine; but a modern council estate would be stretching the imagination rather too far. Equally it should be remembered that very mountainous areas rarely became Napoleonic battlefields; a range of low hills would be much better.

It helps if your terrain can be overlooked from some good vantage point, since this saves a great deal of walking. Alternatively, you can also avoid walking by using an area with plenty of roads, provided that you have transport. Cars are awkward for this purpose, because they will often have to be parked in narrow country lanes. Bicycles are ideal.

Your piece of countryside must have public rights of way across it, if you intend to inspect it from close quarters. It can be infuriating on a TEWT if a crucial part of the battlefield turns out to be private land, and therefore inaccessible.

The size of your area may vary a great deal, and a lot will depend upon the sort of country which is available. A larger area will be needed if it is open country; whereas a lot less will be needed if the horizons are close. In general terms, however, no more than a few square kilometres should be used, and preferably only one or two. Players are supposed to study the ground in detail, so only a small area should be required.

Another factor in selecting the size of your area will be the number of imaginary troops you intend to deploy. The TEWT is best for use with relatively small forces, and a Division should be about the maximum allowed to each side. At the other end of the scale, however, any type of small unit skirmishing is ideally suited for treatment in a TEWT.

All participants should be equipped with a reporter's notebook, or a clip-board and paper. These will allow notes to be made, even in a high wind. Rain will admittedly pose a slight problem, but a TEWT won't be much fun if it's raining, anyway.

Maps may be used on a TEWT, although it is perhaps more realistic to get by without them, as so many Napoleonic officers were forced to do. If there is an umpire, he will need a map more than other players, so an exception should usually be made in his case.

If you are going to conduct a TEWT which covers a lot of ground and takes a long time, you may want a packed lunch at half-time.

Remember, too, that when you are walking in the open air, you should always observe the country code.

Finally, someone must carry a nugget and a small box to roll it in. Open-air dicing can pose some unexpected problems unless an enclosed flat surface is available.

Playing a TEWT

A TEWT is really another form of free kriegspiel: it has no formal rules, and depends entirely upon the common sense and imagination of the players. For guidance, however, something like the following procedure should be used:

Either the umpire or the two leading players should first decide on the scenario, and write a list of the forces on each side, together with their aims. Starting positions and time of day should also be stated, along with any special conditions; e.g., 'that river is in flood', or 'ignore the railway line'.

All the players will now move to the actual ground, and each team will hold a planning conference. At the end of this they should know precisely where to deploy each of their units, and what general plan they want to follow.

Both sides now agree a good place to stand, from which they can overlook the scene of the opening moves. If there is an umpire they should submit their initial moves to him. If there is no umpire, then they tell the other side what movements would first be visible from where they think that side has observers posted. Both sides should therefore accept a little unreality by revealing to the enemy where their advanced scouting screen is posted.

A major difficulty with TEWTs is that there are no markers to show exactly where units have moved, or what formations they have adopted. Players have to make do with just the terrain, and little else apart from their imaginations and memories. Players should therefore make sure that they know exactly what formation each of their units is in, and what each is doing. Written notes will be a very useful aid for this.

As each side notes its moves and announces what the enemy would be able to see, care must be taken to keep everyone informed of everything they would be able to see in reality. If the umpire knows that a particular unit has advanced over a hill, for example, he must announce this to every enemy player who has troops watching that crest line. If there is no umpire, players must simply be honest about this sort of thing, and tell their opponents what would be visible.

Another problem is that it is necessary to make rough estimates for all distances and timings. There is no point in laying down a rigid movement rate, since it will be impossible to measure the distance from one point to another exactly. Players will therefore have to make honest guesses at these things, and ask each other for second opinions. If there is an umpire, he will of course be able to give a more definitive judgement.

A flexible timescale should be used, as far as possible. This means that each turn will represent the time from one important point of decision to the next, rather than any fixed interval of time. Thus the first turn might consist of a brigade of French infantry marching out of a wood and deploying in a field, while the enemy regiment bombards it with artillery. When deployment is complete, a natural stopping place will occur, since the French will then have to decide what to do next and give new orders. The results of the first turn should therefore be assessed at this point.

During the second turn the French might opt to advance across the field and over a hedge. Once again, the crossing of the hedge would make a convenient decision point; so the results of the second turn will be assessed then. In this way the game will progress, with some turns representing a fairly lengthy interval, and others representing a very short one. Once again, it will always be important to keep everyone informed of exactly what is going on.

When there is firing or close combat, the same system is used as in the free kriegspiel. The umpire examines all aspects of the position and quotes odds. The nugget is then rolled, and the final result read off. If there is no umpire, it will be left to the two players concerned to agree on odds between themselves. If no agreement can be reached, then they should each roll a nugget, and the one with the higher score will carry his point.

This type of wargame is obviously very open to cheating and gamesmanship, so it is particularly important to play it with mature and experienced wargamers who will not exploit loopholes. Given the necessary degree of co-operation and understanding, however, it can make a wonderfully different and fascinating exercise. It allows the wargamer to visualize his Napoleonic tactics as they would look on a piece of real ground so he is at last able to escape from the artificial constraints of an indoor game.

Conclusion

The patient reader who has successfully mastered all seven of these games ought by now to have realized that there are indeed more than just one or two approaches to Napoleonic wargaming. There is a different style to suit every mood and every level of action. Provided that the player sets out with a clear idea of precisely what he wants to represent, he should be able to hunt around and come up with a game which pretty closely fills the bill for that occasion.

One final word is called for, about the spirit in which wargames should be played. They are not mathematical exercises in which the players sit silently confronting each other, with their brains ticking over like miniature computers. Instead, they are social events; rather similar to dinner parties. The guest list should be given a little thought in advance, so that all the players will fit into that particular game and enjoy it. The bill of fare should also be designed to suit them; and care should be taken to ensure that no one is left out in the cold. Even the most perfect set of rules (if such a miracle could ever be devised) is useless if it is used on the wrong occasion or with the wrong trimmings; so it is worth while tailoring each game to its social setting. If this is done thoughtfully and well, the result should be an amusing, interesting, and successful evening's entertainment for all concerned.

Glossary of terms and abbreviations

Arty	artillery.
Bde	brigade; a force of several regiments or battalions.
Bn	battalion; a force of several divisions or companies.
Bty	battery; usually six or eight pieces of artillery, with caissons and limbers.
Caisson	a four-wheeled chest for carrying reserve ammunition or other stores.
Deci-dice	(or decimal dice) twenty-sided dice, numbered from 0 to 9 so that they give percentage scores. In this book they are called nuggets.
Division	sub-unit of a corps, composed of several brigades. Always given a capital letter, to distinguish it from:
division	sub-unit of a battalion, usually composed of two companies. We will not use capital letters in this case, to avoid confusion.
Dressing	the alignment of troops, in close-order drill. Units 'take their dressing' when they stop to check and correct their positions in the line.
Gamesmanship	a disease of wargamers who try to exploit the rules at the expense of historical realism and common decency.
Limber	small two-wheeled wagon used for towing cannons and carrying a limited supply of ready-use ammunition. When guns are hitched onto a limber it is called 'limbering up'.
Line of communication	the road by which a force receives supplies and reinforcements from its base, and sends back casualties and prisoners.
Logistics	the science of supplying and quartering troops.
Nugget	a decimal dice (see above).
Rifle	a muzzle loaded flintlock weapon with a twisting groove in the barrel to add spin, and therefore greater range, to the bullet. The common musket, by contrast, had a smooth bore.

Scenario a horrible piece of jargon which has unfortunately made itself indispensable in wargame books. It means 'the hypothetical starting position in the game, including the assumptions from which the players will work'.

TEWT Tactical exercise without troops (see chapter eight).

Appendix

Gamesheet for the skirmish game

Scales 54 mm: 1″ = 1 m; 25 mm: 1″ = 2 m; 1 turn = 10 seconds

Sequence
(i) Write orders
(ii) Moves + morale tests
(iii) Firing
(iv) Close combats
(v) Start next turn

Moves in a turn – foot

Walk:	level	12 m
	rough	10 m
	woods	6 m
Run:	level	18 m
	rough	14 m
	woods	8 m
	0/1 to trip up	
	puffed 1 turn per turn	

Turn more than 90°
Lie/kneel/stand
Open/close door
Start/complete obstacle crossing
Issue an order
Aim + fire (0/1 = misfire)

Firearms except rifles:
- Put in a round
- Ram home (0 = bayonet gash)
- Prime & cock

Rifles:
- Put in round
- Ram home (0 = bayonet gash)
- Get out mallet
- Use mallet
- Prime & cock

Diagnose misfire
Correct misfire (except for 0–3)
Fix/unfix bayonet or draw weapon

Moves in a turn – mounted

Start/complete mounting/dismounting
Hold horses (stop bolts for 8/9 per turn)

	level	*rough*	*woods*
Walk	14	12	6
Trot	20	16	6
Gallop	28	22	(0 = fall)

Jumps: fall for 0–3
Turn horse more than 90°

Wounds
Minor: deduct 1, fall over, test morale
Serious: deduct 3, fall × nuggets score, test morale
2 × serious: die

Morale tests
Within 10 m of enemy
Wound sustained

Roll nugget, fail test if score equal or less than

	score	skill rate
	1	A
	2	B
	3	C
ADD one to skill for cover	4	D
DEDUCT ONE for puffed/minor wound	5	E
DEDUCT TWO for inf. attacked by cav.	6	F
DEDUCT THREE for serious wound	7	G

Firing sequence (i) Check loading (ii) Roll for misfires (iii) Find range (iv) Check number of hits (v) Effect of hits

---

Ranges

	Long	*Effective*
Musket	200	50
Rifle	400	100
Blunderbuss	40	10
Carbine	100	24
Pistol	40	10

Firing hits Roll to equal or exceed:

	LONG RANGE			EFFECTIVE RANGE		
	Target:			*Target*:		
Firer	*Sta.*	*Mov.*	*Cov.*	*Sta.*	*Mov.*	*Cov.*
Composed:						
A/B	7	8	9	5	6	7
C/D	8	9	–	6	7	8
E/F	9	–	–	7	8	9
G	–	–	–	8	9	–
Puffed/mounted/scared:						
A/B	9	9	–	7	8	9
C/D	9	–	–	8	9	–
E/F	–	–	–	9	–	–
G	–	–	–	–	–	–
Two of the above conditions:						
A/B	9	–	–	8	9	–
C/D	9	–	–	9	–	–
E/F/G	–	–	–	–	–	–

All shots at close range hit for a 5

Effect of hits (from fire *or* combat)

0–1: dead
2–5: serious wound
6–9: minor wound

Combat skill

DEDUCT ONE: Prone, facing away, mounted, puffed, scared, enemy higher, enemy armoured or with longer weapon.

DEDUCT TWO: Foot v. mounted walking

DEDUCT THREE: Foot v. mounted trotting/galloping

Combat hits Roll to equal/exceed:

4 skill	A
5	B
6	C
7	D
8	E
9	F

No chance – G

Gamesheet for the divisional game

Scales 1 mm = 1 m; 1 turn = 2 min; 1 figure = 33 men; 1 model gun = 1 bty

Sequence
(i) Decide moves
(ii) Move
(iii) Firing
(iv) Morale tests, combats, initiatives
(v) Rally shaken units
(vi) Orders
(vii) Start next turn

Moves in a turn

Enter/leave house
Mount/dismount
Limber/unlimber
Lie down/stand
Fire (rifle = 2 turns)

or as follows:	*Open*	*Rough*	*Obstacles*
Inf. line	130	90	60
Inf. column	170	110	40
Square/gun by hand	30	10	—
Skirmishers	200	140	80
Guns/vehicles	100	40	—
Horse arty/hy cav	250	200	50
Lt cav./staff	300	250	100
Couriers	350	300	200

(1/2 = $\frac{1}{2}$ rate; 0 = never arrive)

Roads add 10 m

Rough = slopes
ploughland
open wood
rivulet
village

Obstacles = thick wood
hedge/wall etc.
stream
Small river = 4 turns (not vehicles)
Big river = couriers only, drown for 0

Bunching – units within 50 m of each other (rear/side)
Skirmishing – rally with 9 beyond 400 m of main body
Disorganized – prone, bunching, changing formation, crossing obstacles, skirmishing over 400 m

Morale tests
Within 200 m of enemy
Unit tries to pass through
"F" unit moves past
Cav. flank/rear attack; inf. rear attack
Div. commander hit within 200 m
Cavalry which wins a combat may bolt
Roll nugget: fail test for score equal or less than – 1 status A
2 B
3 C
4 D
Rally: roll 9 each turn/combat.

Initiatives in 2nd half of combat turn: roll to equal or exceed

5	A
6	B
7	C
8	D

Ranges

	Max.	*Effective*
Horse arty	750	200
Field arty	1000	200
Heavy arty	1250	200
Musket	200	50
Rifle	400	100
Carbine	100	10

Artillery fire

Firing by		*Target:* *Column*	*Line*	*Cover*
Field	A/B	0.3/0.6	0.2/0.4	0.1/0.2
	C/D	0.2/0.4	0.1/0.2	—/0.1
	E/F	0.1/0.2	—/0.1	—/0.1
Horse	A/B	0.2/0.4	0.1/0.2	—/0.1
	C/D	0.1/0.2	—/0.1	—/0.1
	E/F	—/0.1	—/0.1	—/—
Heavy	A/B	0.5/1.0	0.3/0.6	0.2/0.4
	C/D	0.3/0.6	0.2/0.4	0.1/0.2
	E/F	0.2/0.4	0.1/0.2	0.1/0.1

(Blanks = deduct 0.1 for roll of 7/8/9)

Buildings – fire for 9
Enfilade – tgt upgraded one place
Hard cover – half effect
Guns manhandled ½ turn – fire ½ effect
Individuals – killed for 0, wounded 1/2

Musketry 0.5 status lost if nugget equals/exceeds:

Firer		*Target:* *Formed/open* LONG	SHORT	RANGE	*In cover etc.* LONG	SHORT
Formed	A/B	9	5		–	7
	C/D	9	6		–	8
	E/F	–	8		–	9
Skirmishing	A/B	8	4		9	6
	C/D	9	5		–	7
	E/F	–	8		–	9
Shaken/disorganized		–	8		–	9

Close combat sequence

(i) Skirmishers fall back v. formed unit
(ii) Others halt at 50 m
(iii) Quick rallies
(iv) Crisis initiatives
(v) Add basic unit scores
(vi) Add/deduct factors
(vii) Dice for chance effect
(viii) Find winner and losses
(ix) Does cavalry pursue?

Basic combat scores

	A	B	C	D	E	F
Battery	4	3	2	1	0	0
Battalion	5	4	3	2	1	0
Heavy cav	6	5	4	3	2	1
Light cav	5	4	3	2	1	0

Adjustment to combat score

ADD ONE:	Attacker: Div. commander
	Higher ground: Cover
	Lancers v. inf. and arty
	Armoured cav.: Big unit
ADD TWO:	Hard cover
DEDUCT ONE:	Units which fought last turn
	Defender not in line (inf.)
	Cav. v. inf. column
	Cav. frontal attack on line
	Small unit
DEDUCT TWO:	Cav. on foot
DEDUCT THREE:	Cav. v. square: Unit in col. of route

Effect of chance

Nugget 0 – no change	7 – add 150%
1 – add 25%	8 – add 200%
2 – add 50%	9 – add 400%
3–6 – Add 100%	Defender gets best of draws

Loser withdraw 1 move, shaken; 4–5 lose 1 point
0–1: lose 2 points; 6–7 lose 0.5 points
2–3 lose 1.5 points; 8–9 lose nothing
(less 1 from nugget if different bde in support: less 2 if no support)

Winner: advance 50 m no loss; cav. test morale

Cavalry v. cavalry

(i) Stronger front line pushes back
(ii) 1st line loser tests morale
(iii) All add score as normal, but deduct 1 per front line squadron

Gamesheet for the brigade game

Scales 5 mm = 2 m; 1 turn = 30 seconds; 1 figure = 10 men; 1 model gun = 2 guns

---

Sequence
(i) Write orders (and turn executed)
(ii) Moves and morale tests
(iii) Firing
(iv) Close combat
(v) Start next turn

Orders
Player-figure issues 2 per turn (verbal)
or 1 (written) in 2 turns
Couriers move 120 m; 0/1 = move quarter speed
Bn commanders pass down one order per $\frac{1}{2}$ turn
Company commanders pass down one order per $\frac{1}{2}$ turn
Voice range = 30 m
Signals must be pre-arranged: $\frac{1}{2}$ turn per relay
Initiatives outside orders: for 7/8/9 only

Movement in a turn

Infantry
Walk– 50 good going
30 rough (disorganised)
10 back/side (disorganised)
Run– 70 (Disorganized)
Cross obstacle (disorganized)
Lie/stand/fix/unfix bayonet (disorg.)
Load & fire musket/carbine (disorg.)
Take dressing

Cavalry

	Walk	*Trot*	*Gallop*	
Flat	70	90	100 (hy)	120 (lt)
Rough	50	70	80	

Must accelerate/decelerate walk-trot-gallop
with one turn for each speed
Jump (disorganized): 1–4 less 1 point
0 less 2 points
Mount/dismount

Artillery

	Horse	*Field*
Flat, limbered	70	50
rough, limbered	50	30
Manhandled	20	10

Limber/unlimber/open caisson
Prepare to fire field guns (2 turns hy/horse)
Load + fire (except v. hy – 2 turns)
Traverse more than 90° = $\frac{1}{2}$ turn

Morale tests
First sight of enemy
Any enemy comes within 200 m
Any change of orders for inf. (includes fire)
Unit tries to pass through the section
Bn commander hit within voice range
Neighbouring section fails to rally within 1 turn
In combat

Roll nugget: fail for score of or less than:

Score	Status
1	A
2	B
3	C
4	D

Rally: roll 9

Disorganized units must regain dressing:

Rough ground/obstacles
Skirmishing
Running/galloping
Changing formation
Firing (must also test morale)
All gunners

Ranges

	Max.	*Effective*
Horse arty	750	200
Field arty	1000	200
Heavy arty	1250	200
Musket	200	50
Rifle	400	100
Carbine	100	24

Effect of fire
Arty + 2 on dice
1st volley + 2
Mounted unit − 2
Arty enfilade: throw 2 dice

Roll nugget to equal/exceed

		Column	*Line*	*Cover*
Long range:	A/B	8	9	10
	C/D	9	10	11
	Shaken/E	10	11	12
Effective:	A/B	6	7	8
	C/D	7	8	9
	Shaken/E	8	9	10

Close combat sequence
(i) Agree on units involved
(ii) Roll nugget for interested units
(iii) Find winner and losses

Adjusted nugget roll for sections still interested in the fight

equal/exceed:

	A	B	C	D	E
Formed/in hand	5	6	7	8	9
Disorg./shaken	8	8	9	9	10

Adjustments to combat nugget roll

ADD THREE: Cavalry find flank inf./arty

ADD ONE: Defender uphill/in cover
Attacker uphill/finds flank
Bde commander in voice range

Winner disorganized until takes dressing
all reluctant sections test morale
(all sections which fail lose 1, shaken)

Loser disordered: all test morale
(all which fail lose 1 point, shaken)

Gamesheet for the army level game

Scales 15/25 mm: 1 mm = 1 m; 5 mm: 1 mm = 10 m; one figure = 33 (in 15/25 mm), or one block = 1 bn/bty/cav. regiment (in 5 mm); 1 turn = 15 minutes

---

Sequence
(i) Movement markers placed by units
(ii) Move as previously indicated
(iii) Firing
(iv) Morale tests & close combats
(v) Rally shaken units
(vi) Orders, and start next turn

Movement in a turn

	Open	*Woods*	*Deduct obstacles*
Infantry	700	300	– 200
Arty (hand)	150	50	– 100
Guns/vehicles	500	—	– 450
Horse arty	1200	—	– 1100
Heavy cav.	1200	300	– 500
Lt cav/staff	1500	300	– 500

Couriers: never arrive for 0
Roads – add 150 m
Major rivers – 4 turns, inf/cav.
Big rivers – couriers only, drown for 0
Limbering/unlimbering – $\frac{1}{3}$ turn
Mounting/dismounting – $\frac{1}{3}$ turn

Morale tests
(i) First sight of an enemy
(ii) Unit tries to pass through
(iii) Enemy makes rear attack
(iv) Victorious cavalry may pursue

Roll one nugget, fail test if score equal/less than:

	status
1	A
2	B
3	C
4	D

Rally: roll 8/9 each turn

Ranges

Field bty	– 1000	Rifle	– 400
Horse bty	– 750	Musket	– 200
Heavy bty	– 1250		

Effect of fire per turn: read off status lost by target

Firer	*Target in open*	*Fd bty*	*Horse bty*	*Heavy bty*	*Rifle*	*Musket*
A/B		1.0	0.8	1.2	0.5	0.2
C/D		0.7	0.5	0.9	0.3	0.1
Shaken/E		0.5	0.3	0.7	0.2	—
	cover					
A/B		0.5	0.4	0.6	0.3	0.1
C/D		0.4	0.3	0.5	0.2	—
Shaken/E		0.3	0.2	0.4	0.1	—

Buildings – fired for 7/8/9
Enfilades – double arty score

Close combat All within 200 m of enemy (in woods = skirmishing only)

(i) Find total score of each side	(iii) Find chance variation
(ii) Add/deduct factors	(iv) Find winner & losses

Basic combat scores

A – 5
B – 4
C – 3
D – 2
Skirmish/shaken/E – 1

Combat factors

ADD ONE:	Attacker
	Defender in cover
	Higher ground
	Unit with Div. commander
	Lancers v. inf/arty
	Heavy cav.
ADD TWO:	Defender in hard cover
	Cav. + inf. v. inf.
DEDUCT ONE:	Unit fought last turn
	Caught digging in
	Unit without same bde support
DEDUCT TWO:	Cav. on foot
	Cav. attacking inf.
	Unit in column of route
	Unit without any rear support

Combat chance factor

0 – score unchanged	7 – add 150%
1 – add 25%	8 – add 200%
2 – add 50%	9 – add 400%
3–6 – add 100%	

Winner: cav. test morale

Loser: withdraw, 'shaken',
lose 1 point each

Gamesheet for the generalship game

Scales: Each turn represents 24 hours. The smallest unit is a corps; the map is divided into 'day's marches'.

Types of unit and status

Inf. corps	25	*Supply of towns*:
Res. corps	40	Fortress: 1 corps + 14 days
Cav. corps	10	Open town: 1 corps + 7 days
Grand HQ	–	Siege: 10 × nugget score × days
Little HQ	–	
C.-in-C.'s staff	–	

Possible actions in each turn

Write orders:	$\frac{1}{2}$ hour specify receipt time
Service L. of C. (at LHQ):	$\frac{1}{2}$ hour for base $+\frac{1}{2}$ hour per town
Re-establishing L. of C/HQ:	2 hours
Intelligence (at LHQ):	1 hour per town
Writing home (at LHQ):	1 hour per day or 2 per 2 days: otherwise 4 at start of 3rd day
Letters to other players:	1 hour each
Rest ($\frac{1}{2}$ may be in coach):	as C.-in-C.'s requirement Must take all together if 24 hours' rest is outstanding.
Movement:	Horse, no posts – 3 + 3 Horse with posts – 3 × indefinite (1 extra hour rest per march by horse) Post-chaise – 3 Coach – 4 × indefinite
Inspection:	Hasty $\frac{1}{2}$ hour, 1 status point Formal 4 hours, 4 points
Speech:	$\frac{1}{2}$ hour per corps (1 point)
Personal recce	Foot: 4 hours, 3 points Horse: 2 hours, 2 points Post-chase: 2 hours, 1 point
Battle, in front line:	Killed for 0, wounded for 1/2 add 2 × 3 points, may not write letters etc.

Sequence

(i) Umpires give intelligence available
(ii) Players arrange action counters on gameboard
(iii) Players make notes on orders given
(iv) Umpires check for combats
(v) Players rearrange gameboard after any combat reports
(vi) Umpires assess full result:
Results of combat, effect on moves
All adjustments to status
All adjustments to supply & L. of C.
All expenditure of rest & letters home
(vii) Midnight reports of screen contacts and intelligence gathering
(viii) Clear gameboards & start next turn

Movement rates

Cav.: 6 + 6 (less 1 per forced march)
Res/inf.: 8 + 8 twice a week (less 1 point for any march)
GHQ: 8 (must be 4 marches from res.)
LHQ: 4 + 4
Courier: 4 + 5 × indefinitely
(courier + C.-in-C. horse = cross country)

Combat sequence

(i) Screen reports
(ii) Combat courier report sent
(iii) Both decide posture + reserves
(iv) Is battle immediate, postponed or with 2 hour (withdrawal) delay?
(v) If delayed – C.-in-C.'s morale boosting
(vi) 3 phases = hours × smaller no. of corps
(vii) Find status in each phase:
Fieldwork add 25%
Fortress adds 50%
Unopposed cav. double score
(viii) Nugget roll for chance:
0–5 – no change
6–8 – add 50%
9 – add 100%
(ix) Phase loser – less 20% status
Phase winner – less 10%
(x) Battle ends with 3rd phase, night, or successful withdrawal

---

Siege sequence

(i) Besieger assesses supply: 1 corps indefinitely
(ii) Garrison dices for supply: nugget × 10 days
(iii) Approaches: nugget × 5 days, 0 = abortive
(iv) Breaching: nugget × 3 days, 0 = abortive
(v) Storming: one day – success for 0–5
(vi) Attacker loses total of *all* above dice rolls
Besieged: lose points × 1 nugget (if win)
totally captured (if lose)

Alterations to status

DEDUCT: Each phase of siege or combat
Inf., each march or forced march
Cav., each forced march
Commander hit in battle
Failure of supply that day
Reserve corps too far from GHQ
HQs overrun by enemy

ADD: C.-in-C. in front line, makes recce, makes speech, inspects unit
Every day stationary, add 1 point

Gamesheet for the free kriegspiel and the tactical exercise without troops

Scales Variable (both time and distance). Kriegspiel – start with 1 turn = 24 hours

---

Sequence
(i) Players write orders
(ii) Umpire checks for contacts
(iii) Players submit supplementary contact orders
(iv) Combats are settled and reported
(v) Prepare for next turn

Suggested movement rates (strategic)

	km per day
Infantry	21
Guns/vehicle	21 – on roads only
Heavy cavalry	25
Light cavalry	28
HQ group	31

Couriers 6.5 × 4, then 4 km.p.h. indefinitely

Suggested digging rates (siege)

In 24 hrs – 70 m sap
$\frac{1}{3}$ battery
$\frac{1}{2}$ inf. redoubt
5 m of mine
arm battery/mine

Combat sequence
(i) Umpire checks balance of force, terrain etc.
(ii) Umpire gives probabilities for various possible results
(iii) Nugget roll reveals which applies

Factors for a TEWT
(i) Access to terrain
(ii) Visibility over terrain
(iii) Planned size of forces
(iv) Suitability of terrain features
(v) Foresee possible moves
(vi) Transport
(vii) Notebooks
(viii) Nugget and box
(ix) Maps
(x) Country code
(xi) Intercom between all participants

Further reading

The skirmish game

Other wargame rules

Flintlock and Ramrod, 1700–1850 (published by Skirmish Wargames, 24 Mill Road, Gillingham, Kent ME7 1HN).

General works on Napoleonic troop types

The *Men at Arms* series, (London) [very good for uniforms]
L. & F. Funcken, *Arms and Uniforms – The Napoleonic Wars*, 2 vols (London, 1973).
H.C.B. Rogers, *Napoleon's Army* (London, 1974)

Guides to small group tactics

J.F.C. Fuller, *Sir John Moore's System of Training* (London, 1924)
J. Pimlott, *British Light Cavalry* (London, 1977)
N. de Lee, *French Lancers* (London, 1976)
de Brack, *Outposts of Light Cavalry* (London, 1876)
Harris, *Recollections of Rifleman Harris*, ed. Curling (London, 1929)
P.G. Duhesme, *Essai Historique sur l'Infanterie Legere* (3rd edn, Paris 1864)

Fiction

R.F. Delderfield, *Seven Men of Gascony* (London, 1973)
C.S. Forester, *Death to the French* (New edn, London, 1952)

The Divisional game

Historical

T.A. Dodge, *Napoleon* 4 vols (New York, 1904) (very good for detail of all armies)
Esposito & Elting, *A Military History & Atlas of the Napoleonic Wars* (London, 1964)
J.C. Quennevat, *Atlas de la Grande Armée* (Paris, 1967)
C. Oman, *History of the Peninsular War* 7 vols (Oxford, 1902–30)
D.G. Chandler, *The Campaigns of Napoleon* (London, 1967)
G.E. Rothenberg, *The Art of War in the Age of Napoleon* (London, 1977)

Other wargame rules
D.F. Featherstone, *Wargames* (London, 1962)
B. Quarrie, *Napoleon's Campaigns in Miniature* (Cambridge, 1977)
J. Tunstill, *Discovering Wargaming* (Aylesbury, 1969)
See also many relevant articles in the *Wargamer's Newsletter* and *Military Modelling*

The brigade game

Ardant du Picq, *Battle Studies* (New ed. Harrisburg Pa., 1958)
Colin, *La Tactique et la Discipline dans les Armées de la Révolution* (Paris, 1902)
B.P. Hughes, *Firepower* (London, 1974)
C. Mercer, *Mercer's Journal* (London, 1927)
Thiebault, *Mémoires of Baron Thiebault*, 2 vols. (London, 1896)
Brun, *Cahiers du General Brun* (Paris, 1953); esp. pp. 146–150.

The army level game

H. Cannon, *La Bataille Napoléonienne* (Paris, 1910)
H. Parker, *Three Napoleonic Battles* (London, 1944)
J. Weller, *Wellington at Waterloo* (London, 1967)
C.J. Duffy, *Austerlitz 1805* (London, 1977)

The generalship game

S.G.P. Ward, *Wellington's Headquarters* (Oxford, 1957)
Vachée, *Napoleon at Work* (London, 1914)
G.T. Warner, *How Wars Were Won* (London, 1915)
R. Glover, *Peninsular Preparation – the reform of the British Army 1795–1809* (Cambridge 1963)

Free kriegspiel

Wargames
A. Wilson, *War Gaming* [previously *The Bomb and the Computer*] (London, 1970)

Siegecraft
C.J. Duffy, *Fire and Stone, the Science of Fortress Warfare 1660–1860* (London, 1975)
E. Viollet le Duc, *Annals of a Fortress* (London, 1875)

Imaginary French landings in Britain

E. Desbrière, *Projets et Tentatives de Debarquement aux Iles Britanniques* (Paris, 1900)
I.F. Clarke, *Voices Prophesying War 1763–1984* (Oxford, 1966)
R. Glover, *Britain at Bay* (London, 1973) [British defences against invasion]

Index

Page numbers in italic refer to illustrations